# THE GRAND CANYON

## Early Impressions

# The Grand Canyon: Early Impressions

edited by

## Paul Schullery

Printed in the United States of America

Pruett Publishing Co.
2928 Pearl St.
Boulder, Colorado 80301


Library of Congress Cataloging-in-Publication Data

The Grand Canyon, early impressions / edited by Paul Schullery.
    p.    cm.
Reprint. Originally published: Boulder, Colo.: Colorado Associated University Press, ©198
Bibliography: p.
ISBN 0-87108-765-0
    1. Grand Canyon (Ariz.)—Description and travel.
[F788.G746  1989]
917.91'32044–dc20                            89-3
                                         CI

This book is dedicated to my sister, Debbie Harding

# CONTENTS

# FOREWORD

WE CONTINUE to be fascinated with our past and the views of those who experienced it. This anthology presents a selection of both well-known and obscure essays on the Grand Canyon that date from before the turn of the century to the eve of World War Two. For the repeat visitor or the serious student the editor has performed a valuable service in bringing these long out-of-print words to our attention.

Paul Schullery describes himself as a devoted student of the west with an enthusiastic interest in the history of our great national parks. This volume was inspired by the success of a similar anthology on Yellowstone National Park, and serves as testimony to Mr. Schullery's involvement with and interest in national parks.

The book follows a well-established format in presenting the selections in chronological order. The first and last essays deal with river running, and here we are immersed in the drama of a Colorado River trip and the feelings of those people driven by a need to explore and experience the Canyon from its turbulent waters. Other essays outline the various means of transportation used by determined tourists in reaching the canyon. The first automobile trip to the canyon was naturally planned and carried out by a Californian, but the difficulties and dedication that this four-day trip entailed may come as a surprise to today's travellers. Other stories tell of stage, railroad, mule and airplane trips.

The central portion of the book contains a delightful group of short essays by well-known personalities. Especially enjoyable were the words of John Boynton Priestley, whose message has a special meaning in these troubled times. At the end of the collection are the Sources of the Readings, Notes, and Suggestions for Further Readings that many readers will find most helpful.

It is through readings such as these that we can add a historical perspective to our own experiences, gaining a better understanding and enjoyment. Those visitors who take the time to read these articles will be rewarded with an expanded view of the Grand Canyon.

John C. O'Brien
Executive Secretary
Grand Canyon Natural History Association
Grand Canyon, Arizona

# ACKNOWLEDGMENTS

I WISH to thank Grand Canyon National Park Chief Naturalist John O'Brien and his staff for their assistance in my research. One of the least appreciated services provided by the National Park Service is just this kind of assistance; from my own work as a historian in Yellowstone I know that requests for assistance can be both challenging and time consuming.

The staff of the Mark Skinner Library, Manchester Vermont, must also be mentioned. Though I did my research at several libraries, this one helped frequently with interlibrary loan requests.

Though I do not consider myself a Grand Canyon "authority," I am a devoted student of the American west, and a passionate defender of its parks. My general enthusiasm for national park history and the unique character of the Grand Canyon led me into this project. I first saw the Grand Canyon as a child, and it made a greater impression on my inordinately impressionable mind than any other childhood experience. The Grand Canyon is the only place that I knew as a child that, when I returned as an adult, did not seem smaller. If anything it seemed bigger, as if a child's eyes were less able to take it in than an adult's. Because we shared that memorable first visit, I dedicate this book to my sister.

# Introduction

AMONG the many distinctions that make the Grand Canyon of Arizona unique is the impossibility of its description. No other natural scene has foiled the efforts of so many writers, or been so freely acknowledged as simply indescribable. Artist John McCutcheon suggested that the Grand Canyon was the ultimate challenge for a writer ("one should go into a course of literary training and gradually build up to it"), but he then happily evaded that challenge by telling us to "see other writers" for a description.

The challenge of describing the indescribable has inspired some unusual approaches. We often try to describe by comparison. What does it resemble? "Well," say our early visitors, "it looks a lot like a vast collection of gigantic cathedrals, castles, and towers . . ." And how big is it? "Well, if you stood here on the rim and put the Lower Falls of the Yellowstone River down there at the bottom, nobody would notice." It's just as well for us that writers can not really do what they propose, or by now the Grand Canyon would be overflowing with Statues of Liberty, Empire State Buildings, Queen Marys (end on end), and the ample flow from any number of Niagaras.

The problem is that there is nothing else *like* the Grand Canyon. Not only is there nothing to compare it with, it is beyond comparison:

> No matter how far you have wandered hitherto, or how many famous gorges and valleys you have seen, this one, the Grand Canyon of the Colorado, will seem as novel to you, as unearthly in

1

the color and grandeur and quantity of its architecture, as if you had found it after death, on some other star.[1]

The Grand Canyon was one of many natural wonders being discovered by a growing leisure class at the beginning of the twentieth century. It got a later start than many of them, even those farther west. By 1896, there were already several national parks (including Yosemite and Yellowstone), but it was not until 1908 that Grand Canyon even achieved national monument status. It finally became a national park in 1919. Through all these early years the canyon was being visited, first by a few, then by a steadily increasing crowd of tourists. They arrived in many ways, they enjoyed themselves in many others, and their impressions were as varied as their personalities. But they were all, or mostly all, touched by what they saw, and many of them were moved to share their feelings in writing.

Travel writing, as it developed in the late 1800s, was rarely intended to endure. It differed from narratives of discovery, since it did not describe anything not already visited by others, but it *was* exploratory, in a leisurely way. In order to sell the product, the writer had to discover what subjects appealed to readers, or, more ambitiously, undertake to generate interest in a new place. Among the leading travel writers of the 1880s and 1890s was Charles Dudley Warner, whose account of the canyon is one of the first included here. Warner's biographer, in analyzing Warner's travel books, defined both the purposes and the limitations of travel books in general:

> Paradoxical as it may seem, it is perfectly true that the greatest hindrance to their permanent interest is the information they furnish. The more full, specific and even accurate that is, the more rapidly does the work containing it lose its value. The fresher knowledge conveyed by a new, and it may be much inferior book, crowds out of circulation those which have gone before. The changed or changing conditions in the region traversed renders the information previously furnished out of date and even misleading. Hence the older works come in time to have only an antiquarian interest.[2]

And so the writers who tackled the job of telling about the Grand Canyon seem to have faced a stacked deck. Not only was it common knowledge that it couldn't be done, you could almost expect whatever you wrote to be replaced by something equally unsuccessful right away.

This collection of accounts gives proof of a far greater success and durability than its contributors ever dreamed. In fact, we have here quite a literary triumph. If these selections fall short of bringing the actual wonders of the Grand Canyon into the reader's mind, they are still quite successful in evoking the impressions and awe of visitors there. It is because of *this* success, and the richness and variety these authors bring to the attempt, that this collection of readings is so exciting. After all, the final failure of any writer to do justice to the Grand Canyon is not so much a measure of the writer's inadequacies as it is proof of the magnitude of the task.

The Grand Canyon, so well known as a photographic and artistic subject, has also inspired some fine literature. Some of it is worthy as fine descriptive prose, some as humor. Some has historic value, and some is first rate adventure writing. Some selections are included here simply because they are such splendid "period pieces"; by their transient and dated character they now entertain at least as well as they did when they were current.

No collection of this kind can satisfy everyone. Other choices could have been made, and additional readings are suggested at the end of the book.

The intensity, the warmth, and even the sputtering frustration expressed in these selections all ultimately bear witness to the Grand Canyon's power to reach us, to hold us, and to make us think. However feeble our own efforts to express or define the Grand Canyon experience may be, we, like these authors, are inestimably grateful for the opportunity to try, and are always pleased to return to this special place for renewed inspiration.

# CHAPTER ONE

# 1869
# First Through the
# Grand Canyon

John Wesley Powell

An Australian tourist of my acquaintance once gave me a lecture on "the embarrassing American preoccupation with bigness." Though we are certainly not unique in our pride over size – the tallest building, the longest runway, the deepest canyon – there is in the American attitude an undeniable craving for leadership. It may be just as evident in the passion for being "first": first to fly the Atlantic, first to the Moon, first to neutralize stomach acid, and so on.

Among Grand Canyon enthusiasts, there is surely no more enviable "first" than John Wesley Powell's trip the length of the Colorado River. The undeniable kudos of being first (who can remember the name of the third man to walk on the moon?) has pervaded much of the writing since his expedition. Everyone else has measured their trip against his, carrying his journal with them, quoting – without mercy – evocative passages.

Powell's "first" must be qualified in many ways. He certainly was not the first to see the Grand Canyon. That honor went to some unknown native American thousands of years earlier. He was not even the first person of European origins to see it, since Spanish explorers had preceded him by more than three centuries. Others had even traveled by boat or raft various parts of the Colorado River, and had entered the Grand Canyon itself in their craft. It is not mere historical quirk, though, that John Powell is remembered while the others are largely forgotten. He *was* the first to run the entire river, he was almost certainly the first to run the river the length of the Grand Canyon, and he was surely the first to bring adequate professional descriptions of the inner canyon to the public.

The name of John Wesley Powell is not prominent in most public school history books, but his influence on the development of the American west was probably greater than that of most Presidents. His service in exploring and describing the Colorado River was only the beginning of a distinguished public career (indeed, his public services began even before the exploration; he lost his right arm in 1862, fighting for the Union cause at Shiloh). He was very influential in planning laws to administer the arid western lands, and among his posts were Director of the U.S. Geological Survey (1881-1894) and Head of the Bureau of Ethnology (he died at this position, in 1902). He also helped create the Geological Society of America, and the better known National Geographic Society. He was in many ways an explorer, and a pioneer in America's struggle to manage its vast public domain.

Powell's significance in the exploration and surveying of the west is great, of course, but we should see him, for the purposes of this book, at least, as another kind of pioneer. For all the dangers, and maybe even in spite of his scientific goals, Powell and his men were original tourists. In their response to the canyon's magnificence they differed little from the millions who followed them, millions whose risks were progressively less. Putting aside for a moment Powell's risks and remarkable achievement, the only important difference between him and today's visitor is sequence. Today's canyon traveler, whether river-runner or rim-bound tourist, somewhere during the visit, even if he or she has never heard of Powell, will pay him the homage of envy by wondering "what would it have been like to be *first?*"

A UGUST 13.—We are now ready to start on our way down the Great Unknown. Our boats, tied to a common stake, are chafing each other, as they are tossed by the fretful river. They ride high and buoyant, for their loads are lighter than we could desire. We have but a month's rations remaining. The flour has been resifted through the mosquito net sieve; the spoiled bacon has been dried, and the worst of it boiled; the few pounds of dried apples have been spread in the sun, and reshrunken to their normal bulk; the sugar has all melted, and gone on its way down the river; but we have a large sack of coffee. The lighting of the boats has this advantage: they will ride the waves better, and we shall have but little to carry when we make a portage.

We are three-quarters of a mile in the depths of the earth, and the great river shrinks into insignificance, as it dashes its angry waves against the walls and cliffs, that rise to the world above; they are but puny ripples, and we but pigmies, running up and down the sands, or lost among the boulders.

We have an unknown distance yet to run; an unknown river yet to explore. What falls there are, we know not; what rocks beset the channel, we know not; what walls rise over the river, we know not. Ah, well! we may conjecture many things. The men talk as cheerfully as ever; jests are bandied about freely this morning; but to me the cheer is somber and the jests are ghastly.

With some eagerness, and some anxiety, and some misgiving, we enter the cañon below, and are carried along by the swift water through walls which rise from its very edge. They have the same structure as we noticed yesterday—tiers of irregular shelves below, and, above these, steep slopes to the foot of marble cliffs. We run six miles in a little more than half an hour, and emerge into a more open portion of the cañon, where high hills and ledges of rock intervene between the river and the distant walls. Just at the head of this open place the river runs across a dike; that is, a fissure in the rocks, open to depths below, has been filled with eruptive matter, and this, on cooling, was harder than the rocks through which the crevice was made, and, when these were washed away, the harder volcanic matter remained as a wall, and the river has cut a gate-way through it several hundred feet high, and as many wide. As it crosses the wall, there is a fall below, and a bad rapid,

John Wesley Powell in 1891, twenty-two years after his pioneering trip through the Grand Canyon. *Courtesy of the Grand Canyon Natural History Association*

filled with boulders of trap; so we stop to make a portage. Then on we go, gliding by hills and ledges, with distant walls in view; sweeping past sharp angles of rock; stopping at a few points to examine rapids, which we find can be run, until we have made another five miles, when we land for dinner.

Then we let down with lines, over a long rapid, and start again. Once more the walls close in, and we find ourselves in a narrow gorge, the water again filling the channel, and very swift. With great care, and constant watchfulness, we proceed, making about four miles this afternoon, and camp in a cave.

*August 14.* – At daybreak we walk down the bank of the river, on a little sandy beach, to take a view of a new feature in the

8

cañon. Heretofore, hard rocks have given us bad river; soft rocks, smooth water; and a series of rocks harder than any we have experienced sets in. The river enters the granite!

We can see but a little way into the granite gorge, but it looks threatening.

After breakfast we enter on the waves. At the very introduction, it inspires awe. The cañon is narrower than we have ever before seen it; the water is swifter; there are but few broken rocks in the channel; but the walls are set, on either side, with pinnacles and crags; and sharp, angular buttresses, bristling with wind and wave polished spires, extend far out into the river.

Ledges of rocks jut into the stream, their tops sometimes just below the surface, sometimes rising few or many feet above; and island ledges, and island pinnacles, and island towers break the swift course of the stream into chutes, and eddies, and whirlpools. We soon reach a place where a creek comes in from the left, and just below, the channel is choked with boulders, which have washed down this lateral cañon and formed a dam, over which there is a fall of thirty or forty feet; but on the boulders we can get foothold, and we make a portage.

Three more such dams are found. Over one we make a portage; at the other two we find chutes, through which we can run.

As we proceed, the granite rises higher, until nearly a thousand feet of the lower part of the walls are composed of this rock.

About eleven o'clock we hear a great roar ahead, and approach it very cautiously. The sound grows louder and louder as we run, and at last we find ourselves above a long, broken fall, with ledges and pinnacles of rock obstructing the river. There is a descent of, perhaps, seventy-five or eighty feet in a third of a mile, and the rushing waters break into great waves on the rocks, and lash themselves into a mad, white foam. We can land just above, but there is no foot-hold on either side by which we can make a portage. It is nearly a thousand feet to the top of the granite, so it will be impossible to carry our boats around, though we can climb to the summit up a side gulch, and, passing along a mile or two, can descend to the river. This we find on examination; but such a portage would be impracticable for us, and we must run the rapid, or abandon the river. There is no hesitation. We step into our boats, push off and away we go, first on smooth but swift water, then we strike a glassy wave, and ride to its top, down again into the trough, up again on a higher wave, and down and up on waves higher and still higher, until we strike one just as it curls back, and a breaker rolls over our little boat. Still, on we speed, shooting past projecting rocks, till the little boat is caught in a whirlpool, and spun around several times. At last we pull out again into the stream, and now the other boats have passed us. The open compartment of the *Emma Dean* is filled with water, and every breaker rolls over us. Hurled back from a rock, now on this side, now on that, we are carried into an eddy, in which we struggle for a few minutes, and are then out again, the breakers still rolling over us. Our boat is unmanageable, but she cannot sink, and we drift down another hundred yards, through breakers; how, we scarcely know. We find the other boats have turned into an eddy at the foot of the fall, and are waiting to catch us as we come, for the men have seen that our boat is swamped. They push out as we come near, and pull us in against the wall. We bail our boat, and on we go again.

The walls, now, are more than a mile in height—a vertical distance difficult to appreciate. Stand on the south steps of the Treasury building in Washington, and look down Pennsylvania Avenue to the Capitol Park, and measure this distance overhead, and imagine cliffs to extend to that altitude, and you will understand what I mean; or, stand at Canal Street, in New York, and look up Broadway to Grace Church, and you have about the distance; or, stand at Lake Street bridge, in Chicago, and look down to the Central Depot, and you have it again.

A thousand feet of this is up through granite crags, then steep slopes and perpendicular cliffs rise, one above another, to the summit. The gorge is black and narrow below, red and gray and flaring above, with crags and angular projections on the walls, which, cut in many places by side cañons, seem to be a vast wilderness of rocks. Down in these grand, gloomy depths we glide, ever listening, for the mad waters keep up their roar; ever watching, ever peering ahead, for the narrow cañon is winding, and the river is closed in so that we can see but a few hundred yards, and what there may be below we know not; but we listen for falls, and watch for rocks, or stop now and then, in the bay of a recess, to admire the gigantic scenery. And ever, as we go, there is some new

pinnacle or tower, some crag or peak, some distant view of the
upper plateau, some strange shaped rock, or some deep, narrow
side cañon. Then we come to another broken fall, which appears
more difficult than the one we ran this morning.

A small creek comes in on the right, and the first fall of the
water is over boulders, which have been carried down by this lat-
teral stream. We land at its mouth, and stop for an hour or two to
examine the fall. It seems possible to let down with lines, at least a
part of the way, from point to point, along the right hand wall.
So we make a portage over the first rocks, and find footing on
some boulders below. Then we let down one of the boats to the
end of her line, when she reaches a corner of the projecting rock,
to which one of the men clings, and steadies her, while I examine
an eddy below. I think we can pass the other boats down by us,
and catch them in the eddy. This is soon done and the men in the
boats in the eddy pull us to their side. On the shore of this little
eddy there is about two feet of gravel beach above the water.
Standing on this beach, some of the men take the line of the little
boat and let it drift down against another projecting angle. Here is
a little shelf, on which a man from my boat climbs, and a shorter
line is passed to him, and he fastens the boat to the side of the cliff.
Then the second one is let down, bringing the line of the third.
When the second boat is tied up, the two men standing on the
beach above spring into the last boat, which is pulled up alongside
of ours. Then we let down the boats, for twenty-five or thirty
yards, by walking along the shelf, landing them again in the
mouth of a side cañon. Just below this there is another pile of
boulders, over which we make another portage. From the foot of
these rocks we can climb to another shelf, forty or fifty feet above
the water.

On this beach we camp for the night. We find a few sticks,
which have lodged in the rocks. It is raining hard, and we have no
shelter, but kindle a fire and have our supper. We sit on the rocks
all night, wrapped in our ponchos, getting what sleep we can.

*August 15.* — This morning we find we can let down for three or
four hundred yards, and it is managed in this way: We pass along
the wall, by climbing from projecting point to point, sometimes
near the water's edge, at other places fifty or sixty feet above, and
hold the boat with a line, while two men remain aboard, and

prevent her from being dashed against the rocks, and keep the line from getting caught on the wall. In two hours we have brought them all down, as far as it is possible, in this way. A few yards below, the river strikes with great violence against a projecting rock, and our boats are pulled up in a little bay above. We must now manage to pull out of this, and clear the point below. The little boat is held by the bow obliquely up the stream. We jump in, and pull out only a few strokes, and sweep clear of the dangerous rock. The other boats follow in the same manner, and the rapid is passed.

It is not easy to describe the labor of such navigation. We must prevent the waves from dashing the boats against the cliffs. Sometimes, where the river is swift, we must put a bight of rope about a rock, to prevent her being snatched from us by a wave; but where the plunge is too great, or the chute too swift, we must let her leap, and catch her below, or the undertow will drag her under the falling water, and she sinks. Where we wish to run her out a little way from shore, through a channel between rocks, we first throw in little sticks of drift wood, and watch their course, to see where we must steer, so that she will pass the channel in safety. And so we hold, and let go, and pull, and lift, and ward, among rocks, around rocks, and over rocks.

And now we go on through this solemn, mysterious way. The river is very deep, the cañon very narrow, and still obstructed, so that there is no steady flow of the stream; but the waters wheel, and roll, and boil, and we are scarcely able to determine where we can go. Now, the boat is carried to the right, perhaps close to the wall; again, she is shot into the stream, and perhaps is dragged over to the other side, where, caught in a whirlpool, she spins about. We can neither land nor run as we please. The boats are entirely unmanageable; no order in their running can be preserved; now one, now another, is ahead, each crew laboring for its own preservation. In such a place we come to another rapid. Two of the boats run it perforce. One succeeds in landing, but there is no foot-hold by which to make a portage, and she is pushed out again into the stream. The next minute a great reflex wave fills the open compartment; she is water-logged, and drifts unmanageable. Breaker after breaker rolls over her, and one capsizes her. The men are thrown out; but they cling to the boat, and she drifts down some distance, alongside of us, and we are able to catch her. She is soon bailed out, and the men are aboard once more; but the oars are lost, so a pair from the *Emma Dean* is spared. Then for two miles we find smooth water.

Clouds are playing in the cañon to-day. Sometimes they roll down in great masses, filling the gorge with gloom; sometimes they hang above, from wall to wall, and cover the cañon with a roof of impending storm; and we can peer long distances up and down this cañon corridor, with its cloud roof overhead, its walls of black granite, and its river bright with the sheen of broken waters. Then, a gust of wind sweeps down a side gulch, and, making a rift in the clouds, reveals the blue heavens, and a stream of sunlight pours in. Then, the clouds drift away into the distance, and hang around crags, and peaks, and pinnacles, and towers, and walls, and cover them with a mantle, that lifts from time to time, and sets them all in sharp relief. Then, baby clouds creep out of side cañons, glide around points, and creep back again, into more distant gorges. Then, clouds, set in strata, across the cañon, with intervening vista views, to cliffs and rocks beyond. The clouds are children of the heavens, and when they play among the rocks, they lift them to the region above.

It rains! Rapidly little rills are formed above, and these soon grow into brooks, and the brooks grow into creeks, and tumble over the walls in innumerable cascades, adding their wild music to the roar of the river. When the rain ceases, the rills, brooks, and creeks run dry. The waters that fall, during a rain, on these steep rocks, are gathered at once into the river; they could scarcely be poured in more suddenly, if some vast spout ran from the clouds to the stream itself. When a storm bursts over the cañon, a side gulch is dangerous, for a sudden flood may come, and the in-pouring waters will raise the river, so as to hide the rocks before your eyes.

Early in the afternoon, we discover a stream, entering from the north, a clear, beautiful creek, coming down through a gorgeous red cañon. We land, and camp on a sand beach, above its mouth, under a great, overspreading tree, with willow shaped leaves.

*August 16.* – We must dry our rations again to-day, and make oars.

13

The Colorado is never a clear stream, but for the past three or four days it has been raining much of the time, and the floods, which are poured over the walls, have brought down great quantities of mud, making it exceedingly turbid now. The little affluent, which we have discovered here, is a clear, beautiful creek, or river, as it would be termed in this western country, where streams are not abundant. We have named one stream, away above, in honor of the great chief of the "Bad Angels," and, as this is in beautiful contrast to that, we conclude to name it "Bright Angel."

Early in the morning, the whole party starts up to explore the Bright Angel River, with the special purpose of seeking timber, from which to make oars. A couple of miles above, we find a large pine log, which has been floated down from the plateau, probably from an altitude of more than six thousand feet, but not many miles back. On its way, it must have passed over many cataracts and falls, for it bears scars in evidence of the rough usage which it has received. The men roll it on skids, and the work of sawing oars is commenced.

This stream heads away back, under a line of abrupt cliffs, that terminates the plateau, and tumbles down more than four thousand feet in the first mile or two of its course; then runs through a deep, narrow cañon, until it reaches the river.

Late in the afternoon I return, and go up a little gulch, just above this creek, about two hundred yards from camp, and discover the ruins of two or three old houses, which were originally of stone, laid in mortar. Only the foundations are left, but irregular blocks, of which the houses were constructed, lie scattered about. In one room I find an old mealing stone, deeply worn, as if it had been much used. A great deal of pottery is strewn around, and old trails, which in some places are deeply worn into the rocks, are seen.

It is ever a source of wonder to us why these ancient people sought such inaccessible places for their homes. They were, doubtless, an agricultural race, but there are no lands here, of any considerable extent, that they could have cultivated. To the west of Oraiby, one of the towns in the "Province of Tusayan," in northern Arizona, the inhabitants have actually built little terraces along the face of the cliff, where a spring gushes out, and thus made their sites for gardens. It is possible that the ancient inhabitants of this place made their agricultural lands in the same way. But why should they seek such spots? Surely, the country was not so crowded with population as to demand the utilization of so barren a region. The only solution of the problem suggested is this: We know that, for a century or two after the settlement of Mexico, many expeditions were sent into the country now comprised in Arizona and New Mexico, for the purpose of bringing the town building people under the dominion of the Spanish government. Many of their villages were destroyed, and the inhabitants fled to regions at that time unknown; and there are traditions, among the people who inhabit the *pueblos* that still remain, that the cañons were these unknown lands. Maybe these buildings were erected at that time; sure it is that they have a much more modern appearance than the ruins scattered over Nevada, Utah, Colorado, Arizona, and New Mexico. Those old Spanish conquerers had a monstrous greed for gold, and a wonderful lust for saving souls. Treasures they must have; if not on earth, why, then, in heaven; and when they failed to find heathen temples, bedecked with silver, they propitiated Heaven by seizing the heathen themselves. There is yet extant a copy of a record, made by a heathen artist, to express his conception of the demands of the conquerors. In one part of the picture we have a lake, and near by stands a priest pouring water on the head of a native. On the other side, a poor Indian has a cord about his throat. Lines run from these two groups, to a central figure, a man with beard, and full Spanish panoply. The interpretation of the picture writing is this: "Be baptized, as this saved heathen; or be hanged, as that damned heathen." Doubtless, some of these people preferred a third alternative, and, rather than be baptized or hanged, they chose to be imprisoned within these cañon walls.

*August 17.*—Our rations are still spoiling; the bacon is so badly injured that we are compelled to throw it away. By an accident, this morning, the saleratus is lost overboard. We have now only musty flour sufficient for ten days, a few dried apples, but plenty of coffee. We must make all haste possible. If we meet with difficulties, as we have done in the cañon above, we may be compelled to give up the expedition, and try to reach the Mormon settlements to the north. Our hopes are that the worst places are passed, but our barometers are all so much injured as to be useless, so we have lost our reckoning in altitude, and know not how much descent the river has yet to make.

The stream is still wild and rapid, and rolls through a narrow channel. We make but slow progress, often landing against a wall, and climbing around some point, where we can see the river below. Although very anxious to advance, we are determined to run with great caution, lest, by another accident, we lose all our supplies. How precious that little flour has become! We divide it among the boats, and carefully store it away, so that it can be lost only by the loss of the boat itself.

We make ten miles and a half, and camp among the rocks, on the right. We have had rain, from time to time, all day, and have been thoroughly drenched and chilled; but between showers the sun shines with great power, and the mercury in our thermometers stands at 115°, so that we have rapid changes from great extremes, which are very disagreeable. It is especially cold in the rain to-night. The little canvas we have is rotten and useless; the rubber ponchos, with which we started from Green River City, have all been lost; more than half the party is without hats, and not one of us has an entire suit of clothes, and we have not a blanket apiece. So we gather drift wood, and build a fire; but after supper the rain, coming down in torrents, extinguishes it, and we sit up all night, on the rocks, shivering, and are more exhausted by the night's discomfort than by the day's toil.

*August 18.* – The day is employed in making portages, and we advance but two miles on our journey. Still it rains.

While the men are at work making portages, I climb up the granite to its summit, and go away back over the rust colored sandstones and greenish yellow shales, to the foot of the marble wall. I climb so high that the men and boats are lost in the black depths below, and the dashing river is a rippling brook; and still there is more cañon above than below. All about me are interesting geological records. The book is open, and I can read as I run. All about me are grand views, for the clouds are playing again in the gorges. But somehow I think of the nine days' rations, and the bad river, and the lesson of the rocks, and the glory of the scene is but half seen.

I push on to an angle, where I hope to get a view of the country beyond, to see, if possible, what the prospect may be of our soon running through this plateau, or, at least, of meeting with some geological change that will let us out of the granite; but, arriving at the point, I can see below only a labyrinth of deep gorges.

*August 19.* —Rain again this morning. Still we are in our granite prison, and the time is occupied until noon in making a long, bad portage.

After dinner, in running a rapid, the pioneer boat is upset by a wave. We are some distance in advance of the larger boats, the river is rough and swift, and we are unable to land, but cling to the boat, and are carried down stream, over another rapid. The men in the boats above see our trouble, but they are caught in whirl-pools, and are spinning about in eddies, and it seems a long time before they come to our relief. At last they do come; our boat is turned right side up, bailed out; the oars, which fortunately have floated along in company with us, are gathered up, and on we go, without even landing.

Soon after the accident the clouds break away, and we have sunshine again.

Soon we find a little beach, with just room enough to land. Here we camp, but there is no wood. Across the river, and a little way above, we see some drift wood lodged in the rocks. So we bring two boat loads over, build a huge fire, and spread everything to dry. It is the first cheerful night we have had for a week; a warm, drying fire in the midst of the camp, and a few bright stars in our patch of heavens overhead.

*August 20.* —The characteristics of the cañon change this morning. The river is broader, the walls more sloping, and com-posed of black slates, that stand on edge. These nearly vertical slates are washed out in places—that is, the softer beds are washed out between the harder, which are left standing. In this way, curious little alcoves are formed, in which are quiet bays of water, but on a much smaller scale than the great bays and buttresses of Marble Cañon.

The river is still rapid, and we stop to let down with lines sev-eral times, but make greater progress as we run ten miles. We camp on the right bank. Here, on a terrace of trap, we discover an-other group of ruins. There was evidently quite a village on this

17

rock. Again we find mealing stones, and much broken pottery, and up in a little natural shelf in the rock, back of the ruins, we find a globular basket, that would hold perhaps a third of a bushel. It is badly broken, and, as I attempt to take it up, it falls to pieces. There are many beautiful flint chips, as if this had been the home of an old arrow maker.

*August 21.* – We start early this morning, cheered by the prospect of a fine day, and encouraged, also, by the good run made yesterday. A quarter of a mile below camp the river turns abruptly to the left, and between camp and that point is very swift, running down in a long, broken chute, and piling up against the foot of the cliff, where it turns to the left. We try to pull across, so as to go down on the other side, but the waters are swift, and it seems impossible for us to escape the rock below; but, in pulling across, the bow of the boat is turned to the farther shore, so that we are swept broadside down, and are prevented, by the rebounding waters, from striking against the wall. There we toss about for a few seconds in these billows, and are carried past the danger. Below, the river turns again to the right, the cañon is very narrow, and we see in advance but a short distance. The water, too, is very swift, and there is no landing place. From around this curve there comes a mad roar, and down we are carried, with a dizzying velocity, to the head of another rapid. On either side, high over our heads, there are overhanging granite walls, and the sharp bends cut off our view, so that a few minutes will carry us into unknown waters. Away we go, on one long, winding chute. I stand on deck, supporting myself with a strap, fastened on either side to the gunwale, and the boat glides rapidly, where the water is smooth, or, striking a wave, she leaps and bounds like a thing of life, and we have a wild, exhilarating ride for ten miles, which we make in less than an hour. The excitement is so great that we forget the danger, until we hear the roar of the great fall below; then we back on our oars, and are carried slowly toward its head, and succeed in landing just above, and find that we have to make another portage. At this we are engaged until some time after dinner.

Just here we run out of the granite!

Ten miles in less than half a day, and limestone walls below. Good cheer returns; we forget the storms, and the gloom, and cloud covered cañons, and the black granite, and the raging river, and push our boats from shore in great glee.

Though we are out of the granite, the river is still swift, and we wheel about a point again to the right, and turn, so as to head back in the direction from which we came, and see the granite again, with its narrow gorge and black crags; but we meet with no more great falls, or rapids. Still, we run cautiously, and stop, from time to time, to examine some places which look bad. Yet, we make ten miles this afternoon; twenty miles, in all, to-day.

*August 22.*—We come to rapids again, this morning, and are occupied several hours in passing them, letting the boats down, from rock to rock, with lines, for nearly half a mile, and then have to make a long portage. While the men are engaged in this, I climb the wall on the northeast, to a height of about two thousand five hundred feet, where I can obtain a good view of a long stretch of cañon below. Its course is to the southwest. The walls seem to rise very abruptly, for two thousand five hundred or three thousand feet, and then there is a gently sloping terrace, on each side, for two or three miles, and again we find cliffs, one thousand five hundred or two thousand feet high. From the brink of these the plateau stretches back to the north and south, for a long distance. Away down the cañon, on the right wall, I can see a group of mountains, some of which appear to stand on the brink of the cañon. The effect of the terrace is to give the appearance of a narrow winding valley, with high walls on either side, and a deep, dark, meandering gorge down its middle. It is impossible, from this point of view, to determine whether we have granite at the bottom, or not; but, from geological considerations, I conclude that we shall have marble walls below.

After my return to the boats, we run another mile, and camp for the night.

We have made but little over seven miles to-day, and a part of our flour has been soaked in the river again.

*August 23.*—Our way to-day is again through marble walls. Now and then we pass, for a short distance, through patches of granite, likes hills thrust up into the limestone. At one of these places we have to make another portage, and, taking advantage of the delay, I go up a little stream, to the north, wading it all the way, sometimes having to plunge in to my neck; in other places

being compelled to swim across little basins that have been excavated at the foot of the falls. Along its course are many cascades
and springs gushing out from the rocks on either side. Sometimes
a cottonwood tree grows over the water. I come to one beautiful
fall, of more than a hundred and fifty feet, and climb around it to
the right, on the broken rocks. Still going up, I find the cañon
narrowing very much, being but fifteen or twenty feet wide; yet
the walls rise on either side many hundreds of feet, perhaps thousands; I can hardly tell.

In some places the stream has not excavated its channel down
vertically through the rocks, but has cut obliquely, so that one
wall overhangs the other. In other places it is cut vertically above
and obliquely below, or obliquely above and vertically below, so
that it is impossible to see out overhead. But I can go no farther.
The time which I estimated it would take to make the portage has
almost expired, and I start back on a round trot, wading in the
creek where I must, and plunging through basins, and find the
men waiting for me, and away we go on the river.

Just after dinner we pass a stream on the right, which leaps into
the Colorado by a direct fall of more than a hundred feet, forming
a beautiful cascade. There is a bed of very hard rock above, thirty
or forty feet in thickness, and much softer beds below. The hard
beds above project many yards beyond the softer, which are
washed out, forming a deep cave behind the fall, and the stream
pours through a narrow crevice above into a deep pool below.
Around on the rocks, in the cave like chamber, are set beautiful
ferns, with delicate fronds and enameled stalks. The little frondlets
have their points turned down, to form spore cases. It has very
much the appearance of the Maiden's hair fern, but is much larger.
This delicate foliage covers the rocks all about the fountain, and
gives the chamber great beauty. But we have little time to spend in
admiration, so on we go.

We make fine progress this afternoon, carried along by a swift
river, and shoot over the rapids, finding no serious obstructions.

The cañon walls, for two thousand five hundred or three thousand feet, are very regular, rising almost perpendicularly, but here
and there set with narrow steps, and occasionally we can see away
above the broad terrace, to distant cliffs.

We camp to-night in a marble cave, and find, on looking at our reckoning, we have run twenty-two miles.

*August 24.* —The cañon is wider to-day. The walls rise to a vertical height of nearly three thousand feet. In many places the river runs under a cliff, in great curves, forming amphitheaters, half dome shaped.

Though the river is rapid, we meet with no serious obstructions, and run twenty miles. It is curious how anxious we are to make up our reckoning every time we stop, now that our diet is confined to plenty of coffee, very little spoiled flour, and very few dried apples. It has come to be a race for a dinner. Still, we make such fine progress, all hands are in good cheer, but not a moment of daylight is lost.

*August 25.* —We make twelve miles this morning, when we come to monuments of lava, standing in the river; low rocks, mostly, but some of them shafts more than a hundred feet high. Going on down, three or four miles, we find them increasing in number. Great quantities of cooled lava and many cinder cones are seen on either side; and then we come to an abrupt cataract. Just over the fall, on the right wall, a cinder cone, or extinct volcano, with a well defined crater, stands on the very brink of the cañon. This, doubtless, is the one we saw two or three days ago. From this volcano vast floods of lava have been poured down into the river, and a stream of the molten rock has run up the cañon, three or four miles, and down, we know not how far. Just where it poured over the cañon wall is the fall. The whole north side, as far as we can see, is lined with the black basalt, and high up on the opposite wall are patches of the same material, resting on the benches, and filling old alcoves and caves, giving to the wall a spotted appearance.

The rocks are broken in two, along a line which here crosses the river, and the beds, which we have seen coming down the cañon for the last thirty miles, have dropped 800 feet, on the lower side of the line, forming what geologists call a fault. The volcanic cone stands directly over the fissure thus formed. On the side of the river opposite, mammoth springs burst out of this crevice, one or two hundred feet above the river, pouring in a stream quite equal in volume to the Colorado Chiquito.

21

This stream seems to be loaded with carbonate of lime, and the water, evaporating, leaves an incrustation on the rocks; and this process has been continued for a long time, for extensive deposits are noticed, in which are basins, with bubbling springs. The water is salty.

We have to make a portage here, which is completed in about three hours, and on we go.

We have no difficulty as we float along, and I am able to observe the wonderful phenomena connected with this flood of lava. The cañon was doubtless filled to a height of twelve or fifteen hundred feet, perhaps by more than one flood. This would dam the water back; and in cutting through this great lava bed, a new channel has been formed, sometimes on one side, sometimes on the other. The cooled lava, being of firmer texture than the rocks of which the walls are composed, remains in some places; in others a narrow channel has been cut, leaving a line of basalt on either side. It is possible that the lava cooled faster on the sides against the walls, and that the center ran out; but of this we can only conjecture. There are other places, where almost the whole of the lava is gone, patches of it only being seen where it has caught on the walls. As we float down, we can see that it ran out into side cañons. In some places this basalt has a fine, columnar structure, often in concentric prisms, and masses of these concentric columns have coalesced. In some places, when the flow occurred, the cañon was probably at about the same depth as it is now, for we can see where the basalt has rolled out on the sands, and, what seems curious to me, the sands are not melted or metamorphosed to any appreciable extent. In places the bed of the river is of sandstone or limestone, in other places of lava, showing that it has all been cut out again where the sandstones and limestones appear; but there is a little yet left where the bed is of lava.

What a conflict of water and fire there must have been here! Just imagine a river of molten rock, running down into a river of melted snow. What a seething and boiling of the waters; what clouds of steam rolled into the heavens!

Thirty-five miles to-day. Hurrah!

*August 26.* —The cañon walls are steadily becoming higher as we advance. They are still bold, and nearly vertical up to the terrace. We still see evidence of the eruption discovered yesterday, but the thickness of the basalt is decreasing, as we go down the stream; yet it has been reinforced at points by streams that have come down from volcanoes standing on the terrace above, but which we cannot see from the river below.

Since we left the Colorado Chiquito, we have seen no evidences that the tribe of Indians inhabiting the plateaus on either side ever come down to the river; but about eleven o'clock to-day we discover an Indian garden, at the foot of the wall on the right, just where a little stream, with a narrow flood plain, comes down through a side cañon. Along the valley, the Indians have planted corn, using the water which burst out in springs at the foot of the cliff, for irrigation. The corn is looking quite well, but is not sufficiently advanced to give us roasting ears; but there are some nice, green squashes. We carry ten or a dozen of these on board our boats, and hurriedly leave, not willing to be caught in the robbery, yet excusing ourselves by pleading our great want. We run down a short distance, to where we feel certain no Indians can follow; and what a kettle of squash sauce we make! True, we have no salt with which to season it, but it makes a fine addition to our unleavened bread and coffee. Never was fruit so sweet as these stolen squashes.

After dinner we push on again, making fine time, finding many rapids, but none so bad that we cannot run them with safety, and when we stop, just at dusk, and foot up our reckoning, we find we have run thirty-five miles again.

What a supper we make; unleavened bread, green squash sauce, and strong coffee. We have been for a few days on half rations, but we have no stint of roast squash.

A few days like this, and we shall be out of prison.

*August 27.*—This morning the river takes a more southerly direction. The dip of the rocks is to the north, and we are rapidly running into lower formations. Unless our course changes, we shall very soon run again into the granite. This gives us some anxiety. Now and then the river turns to the west, and excites hopes that are soon destroyed by another turn to the south. About nine o'clock we come to the dreaded rock. It is with no little misgiving that we see the river enter these black, hard walls. At its very entrance we have to make a portage; then we have to let down with lines past some ugly rocks. Then we run a mile or two farther, and then the rapids below can be seen.

About eleven o'clock we come to a place in the river where it seems much worse than any we have yet met in all its course. A little creek comes down from the left. We land first on the right, and clamber up over the granite pinnacles for a mile or two, but can see no way by which we can let down, and to run it would be sure destruction. After dinner we cross to examine it on the left. High above the river we can walk along on the top of the granite, which is broken off at the edge, and set with crags and pinnacles, so that it is very difficult to get a view of the river at all. In my eagerness to reach a point where I can see the roaring fall below, I go too far on the wall, and can neither advance nor retreat. I stand with one foot on a little projecting rock, and cling with my hand fixed in a little crevice. Finding I am caught here, suspended 400 feet above the river, into which I should fall if my footing fails, I call for help. The men come, and pass me a line, but I cannot let go of the rock long enough to take hold of it.* Then they bring two or three of the largest oars. All this takes time which seems very precious to me; but at last they arrive. The blade of one of the oars is pushed into a little crevice in the rock beyond me, in such a manner that they can hold me pressed against the wall. Then another is fixed in such a way that I can step on it, and thus I am extricated.

Still another hour is spent in examining the river from this side, but no good view of it is obtained, so now we return to the side that was first examined, and the afternoon is spent in clambering among the crags and pinnacles, and carefully scanning the river again. We find that the lateral streams have washed boulders into the river, so as to form a dam, over which the water makes a broken fall of eighteen or twenty feet; then there is a rapid, beset with rocks, for two or three hundred yards, while, on the other side, points of the wall project into the river. Then there is a second fall below; how great, we cannot tell. Then there is a rapid, filled with huge rocks, for one or two hundred yards. At the bottom of it, from the right wall, a great rock projects quite half way across the river. It has a sloping surface extending up stream,

---

*It should be remembered that Major Powell had only one arm. (*Ed.*)

24

and the water, coming down with all the momentum gained in the falls and rapids above, rolls up this inclined plane many feet, and tumbles over to the left. I decide that it is possible to let down over the first fall, then run near the right cliff to a point just above the second, where we can pull out into a little chute, and, having run over that in safety, we must pull with all our power across the stream, to avoid the great rock below. On my return to the boat, I announce to the men that we are to run it in the morning. Then we cross the river, and go into camp for the night on some rocks, in the mouth of the little cañon.

After supper Captain Howland asks to have a talk with me. We walk up the little creek a short distance, and I soon find that his object is to remonstrate against my determination to proceed. He thinks that we had better abandon the river here. Talking with him, I learn that his brother, William Dunn, and himself have determined to go no farther in the boats. So we return to camp. Nothing is said to the other men.

For the last two days, our course has not been plotted. I sit down and do this now, for the purpose of finding where we are by dead reckoning. It is a clear night, and I take out the sextant to make observation for latitude, and find that the astronomic determination agrees very nearly with that of the plot—quite as closely as might be expected, from a meridian observation on a planet. In a direct line, we must be about forty-five miles from the mouth of the Rio Virgen. If we can reach that point, we know that there are settlements up that river about twenty miles. This forty-five miles, in a direct line, will probably be eighty or ninety in the meandering line of the river. But then we know that there is comparatively open country for many miles above the mouth of the Virgen, which is our point of destination.

As soon as I determine all this, I spread my plot on the sand, and wake Howland, who is sleeping down by the river, and show him where I suppose we are, and where several Mormon settlements are situated.

We have another short talk about the morrow, and he lies down again; but for me there is no sleep. All night long, I pace up and down a little path, on a few yards of sand beach, along by the river. Is it wise to go on? I go to the boats again, to look at our rations. I feel satisfied that we can get over the danger immediately before us; what there may be below I know not. From our outlook yesterday, on the cliffs, the cañon seemed to make another great bend to the south, and this, from our experience heretofore, means more and higher granite walls. I am not sure that we can climb out of the cañon here, and, when at the top of the wall, I know enough of the country to be certain that it is a desert of rock and sand, between this and the nearest Mormon town, which, on the most direct line, must be seventy-five miles away. True, the late rains have been favorable to us, should we go out, for the probabilities are that we shall find water still standing in holes, and, at one time, I almost conclude to leave the river. But for years I have been contemplating this trip. To leave the exploration unfinished, to say that there is a part of the cañon which I cannot explore, having already almost accomplished it, is more than I am willing to acknowledge, and I determine to go on.

I wake my brother, and tell him of Howland's determination, and he promises to stay with me; then I call up Hawkins, the cook, and he makes a like promise; then Sumner, and Bradley, and Hall, and they all agree to go on.

*August 28.* – At last daylight comes, and we have breakfast, without a word being said about the future. The meal is as solemn as a funeral. After breakfast, I ask the three men if they still think it best to leave us. The elder Howland thinks it is, and Dunn agrees with him. The younger Howland tries to persuade them to go on with the party, failing in which, he decides to go with his brother.

Then we cross the river. The small boat is very much disabled, and unseaworthy. With the loss of hands, consequent on the departure of the three men, we shall not be able to run all of the boats, so I decide to leave my *Emma Dean.*

Two rifles and a shotgun are given to the men who are going out. I ask them to help themselves to the rations, and take what they think to be a fair share. This they refuse to do, saying they have no fear but that they can get something to eat; but Billy, the cook, has a pan of biscuits prepared for dinner, and these he leaves on a rock.

Before starting, we take our barometers, fossils, the minerals, and some ammunition from the boat, and leave them on the rocks. We are going over this place as light as possible. The three men help us lift our boats over a rock twenty-five or thirty feet high, and let them down again over the first fall, and now we are all ready to start. The last thing before leaving, I write a letter to my wife, and give it to Howland. Sumner gives him his watch, directing that it be sent to his sister, should he not be heard from again. The records of the expedition have been kept in duplicate. One set of these is given to Howland, and now we are ready. For the last time, they entreat us not to go on, and tell us that it is madness to set out in this place; that we can never get safely through it; and, further, that the river turns again to the south into the granite, and a few miles of such rapids and falls will exhaust our entire stock of rations, and then it will be too late to climb out. Some tears are shed; it is rather a solemn parting; each party thinks the other is taking the dangerous course.

My old boat left, I go on board of the *Maid of the Cañon*. The three men climb a crag, that overhangs the river, to watch us off. The *Maid of the Cañon* pushes out. We glide rapidly along the foot of the wall, just grazing one great rock, then pull out a little into the chute of the second fall, and plunge over it. The open compartment is filled when we strike the first wave below, but we cut through it, and then the men pull with all their power toward the left wall, and swing clear of the dangerous rock below all right. We are scarcely a minute in running it, and find that, although it looked bad from above, we have passed many places that were worse.

The other boat follows without more difficulty. We land at the first practicable point below and fire our guns, as a signal to the men above that we have come over in safety. Here we remain a couple of hours, hoping that they will take the smaller boat and follow us. We are behind a curve in the cañon, and cannot see up to where we left them, and so we wait until their coming seems hopeless, and push on.

And now we have a succession of rapids and falls until noon, all of which we run in safety. Just after dinner we come to another bad place. A little stream comes in from the left, and below there is a fall, and still below another fall. Above, the river tumbles down, over and among the rocks, in whirlpools and great waves, and the waters are lashed into mad, white foam. We run along the left, above this, and soon see that we cannot get down on this side, but it seems possible to let down on the other. We pull up stream

27

again, for two or three hundred yards, and cross. Now there is a bed of basalt on this northern side of the cañon, with a bold escarpment, that seems to be a hundred feet high. We can climb it, and walk along its summit to a point where we are just at the head of the fall. Here the basalt is broken down again, so it seems to us, and I direct the men to take a line to the top of the cliff, and let the boats down along the wall. One man remains in the boat, to keep her clear of the rocks, and prevent her line from being caught on the projecting angles. I climb the cliff, and pass along to a point just over the fall, and descend by broken rocks, and find that the break of the fall is above the break of the wall, so that we cannot land; and that still below the river is very bad, and that there is no possibility of a portage.

Without waiting further to examine and determine what shall be done, I hasten back to the top of the cliff, to stop the boats from coming down. When I arrive, I find the men have let one of them down to the head of the fall. She is in swift water, and they are not able to pull her back; nor are they able to go on with the line, as it is not long enough to reach the higher part of the cliff, which is just before them; so they take a bight around a crag. I send two men back for the other line. The boat is in very swift water, and Bradley is standing in the open compartment, holding out his oar to prevent her from striking against the foot of the cliff. Now she shoots out into the stream, and up as far as the line will permit, and then, wheeling, drives headlong against the rock, then out and back again, now straining on the line, now striking against the rock. As soon as the second line is brought, we pass it down to him; but his attention is all taken up with his own situation, and he does not see that we are passing the line to him. I stand on a projecting rock, waving my hat to gain his attention, for my voice is drowned by the roaring of the falls.

Just at this moment, I see him take his knife from its sheath, and step forward to cut the line. He has evidently decided that it is better to go over with the boat as it is, than to wait for her to be broken to pieces. As he leans over, the boat sheers again into the stream, the stem-post breaks away, and she is loose. With perfect composure Bradley seizes the great scull oar, places it in the stern rowlock, and pulls with all his power (and he is an athlete) to turn the bow of the boat down stream, for he wishes to go bow down, rather than to drift broadside on. One, two strokes he makes, and a third just as she goes over, and the boat is fairly turned, and she goes down almost beyond our sight, though we are more than a hundred feet above the river. Then she comes up again, on a great wave, and down and up, then around behind some great rocks, and is lost in the mad, white foam below. We stand frozen with fear, for we see no boat. Bradley is gone, so it seems. But now, away below, we see something coming out of the waves. It is evidently a boat. A moment more, and we see Bradley standing on deck, swinging his hat to show that he is all right. But he is in a whirlpool. We have the stem-post of his boat attached to the line. How badly she may be disabled we know not.

I direct Sumner and Powell to pass along the cliff, and see if they can reach him from below. Rhodes, Hall, and myself run to the other boat, jump aboard, push out, and away we go over the falls. A wave rolls over us and our boat is unmanageable. Another great wave strikes us, the boat rolls over, and tumbles and tosses, I know not how. All I know is that Bradley is picking us up. We soon have all right again, and row to the cliff, and wait until Sumner and Powell can come. After a difficult climb they reach us. We run two or three miles farther, and turn again to the northwest, continuing until night, when we have run out of the granite once more.

*August 29.* —We start very early this morning. The river still continues swift, but we have no serious difficulty, and at twelve o'clock emerge from the Grand Cañon of the Colorado. We are in a valley now, and low mountains are seen in the distance, coming to the river below. We recognize this as the Grand Wash.

A few years ago, a party of Mormons set out from St. George, Utah, taking with them a boat, and came down to the mouth of the Grand Wash, where they divided, a portion of the party crossing the river to explore the San Francisco Mountains. Three men—Hamblin, Miller, and Crosby—taking the boat, went on down the river to Callville, landing a few miles below the mouth of the Rio Virgen. We have their manuscript journal with us, and so the stream is comparatively well known.

To-night we camp on the left bank, in a *mesquite* thicket.

The relief from danger, and the joy of success, are great. When he who has been chained by wounds to a hospital cot, until his canvas tent seems like a dungeon cell, until the groans of those who lie about, tortured with probe and knife, are piled up, a weight of horror on his ears that he cannot throw off, cannot forget, and until the stench of festering wounds and anaesthetic drugs has filled the air with its loathsome burthen, at last goes out into the open field, what a world he sees! How beautiful the sky; how bright the sunshine; what "floods of delirious music" pour from the throats of birds; how sweet the fragrance of earth, and tree, and blossom! The first hour of convalescent freedom seems rich recompense for all—pain, gloom, terror.

Something like this are the feelings we experience to-night. Ever before us has been an unknown danger, heavier than immediate peril. Every waking hour passed in the Grand Cañon has been one of toil. We have watched with deep solicitude the steady disappearance of our scant supply of rations, and from time to time have seen the river snatch a portion of the little left, while we were ahungered. And danger and toil were endured in those gloomy depths, where ofttimes the clouds hid the sky by day, and but a narrow zone of stars could be seen at night. Only during the few hours of deep sleep, consequent on hard labor, has the roar of the waters been hushed. Now the danger is over; now the toil has ceased; now the gloom has disappeared; now the firmament is bounded only by the horizon; and what a vast expanse of constellations can be seen!

The river rolls by us in silent majesty; the quiet of the camp is sweet; our joy is almost ecstasy. We sit till long after midnight, talking of the Grand Cañon, talking of home, but chiefly talking of the three men who left us. Are they wandering in those depths, unable to find a way out? are they searching over the desert lands above for water? or are they nearing the settlements?

*August 30.*—We run through two or three short, low cañons to-day, and on emerging from one, we discover a band of Indians in the valley below. They see us, and scamper away in most eager haste, to hide among the rocks. Although we land, and call for them to return, not an Indian can be seen.

Two or three miles farther down, in turning a short bend in the river, we come upon another camp. So near are we before they can see us that I can shout to them, and, being able to speak a little of their language, I tell them we are friends; but they all flee to the rocks, except a man, a woman, and two children. We land, and talk with them. They are without lodges, but have built little shelters of boughs, under which they wallow in the sand. The man is dressed in a hat; the woman in a string of beads only. At first they are evidently much terrified; but when I talk to them in their own language, and tell them we are friends, and inquire after people in the Mormon towns, they are soon reassured, and beg for tobacco. Of this precious article we have none to spare. Sumner looks around in the boat for something to give them, and finds a little piece of colored soap, which they receive as a valuable present, rather as a thing of beauty than as a useful commodity, however. They are either unwilling or unable to tell us anything about the Indians or white people, and so we push off, for we must lose no time.

We camp at noon under the right bank. And now, as we push out, we are in great expectancy, for we hope every minute to discover the mouth of the Rio Virgen.

Soon one of the men explains: "Yonder's an Indian in the river." Looking for a few minutes, we certainly do see two or three persons. The men bend to their oars, and pull toward them. Approaching, we see that there are three white men and an Indian hauling a seine, and then we discover that it is just at the mouth of the long sought river.

As we come near, the men seem far less surprised to see us than we do to see them. They evidently know who we are, and, on talking with them, they tell us that we have been reported lost long ago, and that some weeks before, a messenger had been sent from Salt Lake City, with instructions for them to watch for any fragments or relics of our party that might drift down the stream.

Our new found friends, Mr. Asa and his two sons, tell us that they are pioneers of a town that is to be built on the bank.

Eighteen or twenty miles up the valley of the Rio Virgen there are two Mormon towns, St. Joseph and St. Thomas. To-night we dispatch an Indian to the last mentioned place, to bring any letters that may be there for us.

Our arrival here is very opportune. When we look over our store of supplies, we find about ten pounds of flour, fifteen pounds of dried apples, but seventy or eighty pounds of coffee.

*August 31.*—This afternoon the Indian returns with a letter, informing us that Bishop Leithhead, of St. Thomas, and two or three other Mormons are coming down with a wagon, bringing us supplies. They arrive about sundown. Mr. Asa treats us with great kindness, to the extent of his ability; but Bishop Leithhead brings in his wagon two or three dozen melons, and many other little luxuries, and we are comfortable once more.

*September 1.*—This morning Sumner, Bradley, Hawkins, and Hall, taking on a small supply of rations, start down the Colorado with the boats. It is their intention to go to Fort Mojave, and perhaps from there overland to Los Angeles.

Captain Powell and myself return with Bishop Leithhead to St. Thomas. From St. Thomas we go to Salt Lake City.

CHAPTER TWO

# 1891
# The Heart of the Desert

## Charles Dudley Warner

Charles Dudley Warner was a popular essayist in the late Nineteenth Century, one of his specialties being travel writing. He
wrote many books, both fiction and nonfiction, and was a con-
tributing editor to *Harper's Magazine* from 1886 to 1898.[1]

Warner has been described as the "first noted author to tell
the world" about the canyon.[2] Canyon tourism was in its
infancy at the time of his visit (one source says the first true
tourist reached the canyon in 1884[3]), mostly because the area
was so inaccessible. An added appeal of these earliest accounts
of the canyon is their sense of adventure: the trip *to* the canyon
was almost as memorable as the canyon itself. Added to the
excitement of the journey—then as now—was the surprise and
delight of discovering the distinctive cultural elements of the
American Southwest. As often as we set off to visit some
famous natural wonder our visit is enriched by the human interest of the place. Warner's passing references to Indians,
cliff-dwellings, and cowboy outposts stir our imaginations and
remind us that the geologic forces that shaped the canyon also
shaped its inhabitants.

T HERE is an arid region lying in Northern Arizona and Southern Utah which has been called the District of the Grand Cañon of the Colorado. The area, roughly estimated, contains from 13,000 to 16,000 square miles—about the size of the state of Maryland. This region, fully described by the explorers and studied by the geologists in the United States service, but little known to even the travelling public, is probably the most interesting territory of its size on the globe. At least it is unique. In attempting to convey an idea of it the writer can be assisted by no comparison, nor can he appeal in the minds of his readers to any experience of scenery that can apply here. The so-called Grand Cañon differs not in degree from all other scenes; it differs in kind.

The Colorado River flows southward through Utah, and crosses the Arizona line below the junction with the San Juan. It continues southward, flowing deep in what is called the Marble Cañon, till it is joined by the Little Colorado, coming up from the southeast; it then turns westward in a devious line until it drops straight south, and forms the western boundary of Arizona. The center of the district mentioned is the westwardly flowing part of the Colorado. South of the river is the Colorado Plateau, at a general elevation of about 7000 feet. North of it the land is higher, and ascends in a series of plateaus, and then terraces, a succession of cliffs like a great stairway, rising to the high plateaus of Utah. The plateaus, adjoining the river on the north and well marked by north and south dividing lines, or faults, are, naming them from east to west, the Paria, the Kaibab, the Kanab, the Uinkaret, and the Sheavwitz, terminating in a great wall on the west, the Great Wash fault, where the surface of the country drops at once from a general elevation of 6000 feet to from 1300 to 3000 feet above the sea-level—into a desolate and formidable desert.

If the Grand Cañon itself did not dwarf everything else, the scenery of these plateaus would be superlative in interest. It is not all desert, nor are the gorges, cañons, cliffs, and terraces, which gradually prepare the mind for the comprehension of the Grand Cañon, the only wonders of this land of enchantment. These are contrasted with the sylvan scenery of the Kaibab Plateau, its giant forests and parks, and broad meadows decked in the summer with

34

wild flowers in dense masses of scarlet, white, purple, and yellow. The Vermilion Cliffs, the Pink Cliffs, the White Cliffs, surpass in fantastic form and brilliant color anything that the imagination conceives possible in nature, and there are dreamy landscapes quite beyond the most exquisite fancies of Claude and of Turner. The region is full of wonders, of beauties, and sublimities that Shelley's imaginings do not match in the "Prometheus Unbound," and when it becomes accessible to the tourist it will offer an endless field for the delight of those whose minds can rise to the heights of the sublime and the beautiful. In all imaginative writing or painting the material used is that of human experience, otherwise it could not be understood; even heaven must be described in the terms of an earthly paradise. Human experience has no prototype of this region, and the imagination has never conceived of its forms and colors. It is impossible to convey an adequate idea of it by pen or pencil or brush. The reader who is familiar with the glowing descriptions in the official reports of Major J. W. Powell, Captain C. E. Dutton, Lieutenant Ives, and others, will not save himself from a shock of surprise when the reality is before him. This paper deals only with a single view in this marvellous region.

The point where we struck the Grand Cañon, approaching it from the south, is opposite the promontory in the Kaibab Plateau named Point Sublime by Major Powell, just north of the 36th parallel, and 112° 15′ west longitude. This is only a few miles west of the junction with the Little Colorado. About three or four miles west of this junction the river enters the east slope of the east Kaibab monocline, and here the Grand Cañon begins. Rapidly the chasm deepens to about 6000 feet, or rather it penetrates a higher country, the slope of the river remaining about the same. Through this lofty plateau—an elevation of 7000 to 9000 feet—the chasm extends for sixty miles, gradually changing its course to the north-west, and entering the Kanab Plateau. The Kaibab division of the Grand Cañon is by far the sublimest of all, being 1000 feet deeper than any other. It is not grander only on account of its greater depth, but it is broader and more diversified with magnificent architectural features.

The Kanab division, only less magnificent than the Kaibab, receives the Kanab Cañon from the north and the Cataract Cañon from the south, and ends at the Toroweap Valley.

The section of the Grand Cañon seen by those who take the route from Peach Springs is between 113° and 114° west longitude, and, though wonderful, presents few of the great features of either the Kaibab or the Kanab divisions. The Grand Cañon ends, west longitude 114°, at the Great Wash, west of the Hurricane Ledge or Fault. Its whole length from Little Colorado to the Great Wash, measured by the meanderings of the surface of the river, is 220 miles; by a median line between the crests of the summits of the walls with two-mile cords, about 195 miles; the distance in a straight line is 125 miles.

In our journey to the Grand Cañon we left the Santa Fé line at Flagstaff, a new town with a lively lumber industry, in the midst of a spruce-pine forest which occupies the broken country through which the road passes for over fifty miles. The forest is open, the trees of moderate size are too thickly set with low-growing limbs to make clean lumber, and the foliage furnishes the minimum of shade; but the change to these woods is a welcome one from the treeless reaches of the desert on either side. The cañon is also reached from Williams, the next station west, the distance being a little shorter, and the point on the cañon visited being usually a little farther west. But the Flagstaff route is for many reasons usually preferred. Flagstaff lies just south-east of the San Francisco Mountain, and on the great Colorado Plateau, which has a pretty uniform elevation of about 7000 feet above the sea. The whole region is full of interest. Some of the most remarkable cliff dwellings are within ten miles of Flagstaff, on the Walnut Creek Cañon. At Holbrook, 100 miles east, the traveller finds a road some forty miles long, that leads to the great petrified forest, or Chalcedony Park. Still farther east are the villages of the Pueblo Indians, near the line, while to the northward is the great reservation of the Navajos, a nomadic tribe celebrated for its fine blankets and pretty work in silver—a tribe that preserves much of its manly independence by shunning the charity of the United States. No Indians have come into intimate or dependent relations with the whites without being deteriorated.

Flagstaff is the best present point of departure, because it has a small hotel, good supply stores, and a large livery-stable, made necessary by the business of the place and the objects of interest in the neighborhood, and because one reaches from there by the easiest road the finest scenery incomparably on the Colorado. The distance is seventy-six miles through a practically uninhabited country, much of it a desert, and with water very infrequent. No work has been done on the road; it is made simply by driving over it. There are a few miles here and there of fair wheeling, but a good deal of it is intolerably dusty or exceedingly stony, and progress is slow. In the daytime (it was the last of June) the heat is apt to be excessive; but this could be borne, the air is so absolutely dry and delicious, and breezes occasionally spring up, if it were not for the dust. It is, notwithstanding the novelty of the adventure and of the scenery by the way, a tiresome journey of two days. A day of rest is absolutely required at the cañon, so that five days must be allowed for the trip. This will cost the traveller, according to the size of the party made up, from forty to fifty dollars. But a much longer sojourn at the cañon is desirable.

Our party of seven was stowed in and on an old Concord coach drawn by six horses, and piled with camp equipage, bedding, and provisions. A four-horse team followed, loaded with other supplies and cooking utensils. The road lies on the east side of the San Francisco Mountain. Returning, we passed around its west side, gaining thus a complete view of this shapely peak. The compact range is a group of extinct volcanoes, the craters of which are distinctly visible. The cup-like summit of the highest is 13,000 feet above the sea, and snow always lies on the north escarpment. Rising about 6000 feet above the point of view of the great plateau, it is from all sides a noble object, the dark rock, snow-sprinkled, rising out of the dense growth of pine and cedar. We drove at first through open pine forests, through park-like intervals, over the foot-hills of the mountain, through growths of scrub cedar, and out into the ever-varying rolling country to widely-extended prospects. Two considerable hills on our right attracted us by their unique beauty. Upon the summit and side of each was a red glow exactly like the tint of sunset. We thought surely that it was the effect of reflected light, but the sky was cloudless and the color remained constant. The color came from the soil. The first was called Sunset Mountain. One of our party named the other, and the more beautiful, Peachblow Mountain, a poetic and perfectly descriptive name.

We lunched at noon beside a swift, clouded, cold stream of snow-water from the San Francisco, along which grew a few gnarled cedars and some brilliant wild flowers. The scene was more than picturesque; in the clear hot air of the desert the distant landscape made a hundred pictures of beauty. Behind us the dark form of San Francisco rose up 6000 feet to its black crater and fields of spotless snow. Away off to the northeast, beyond the brown and gray pastures, across a far line distinct in dull color, lay the Painted Desert, like a mirage, like a really painted landscape, glowing in red and orange and pink, an immense city rather than a landscape, with towers and terraces and façades, melting into indistinctness as in a rosy mist, spectral but constant, weltering in a tropic glow and heat, walls and columns and shafts, the wreck of an Oriental capital on a wide violet plain, suffused with brilliant color softened into exquisite shades. All over this region nature has such surprises, that laugh at our inadequate conception of her resources.

Our camp for the night was at the next place where water could be obtained, a station of the Arizona Cattle Company. Abundant water is piped down to it from mountain springs. The log-house and stable of the cowboys were unoccupied, and we pitched our tent on a knoll by the corral. The night was absolutely dry, and sparkling with the starlight. A part of the company spread their blankets on the ground under the sky. It is apt to be cold in this region towards morning, but lodging in the open air is no hardship in this delicious climate. The next day the way part of the distance, with only a road marked by wagon wheels, was through extensive and barren-looking cattle ranges, through pretty vales of grass surrounded by stunted cedars, and over stormy ridges and plains of sand and small boulders. The water having failed at Red Horse, the only place where it is usually found in the day's march, our horses went without, and we had resource to our canteens. The whole country is essentially arid, but snow falls in the wintertime, and its melting, with occasional showers in the summer, create what are called surface wells, made by drainage. Many of them go dry by June. There had been no rain in the region since the last of March, but clouds were gathering daily, and showers are always expected in July. The phenomenon of rain on this baked surface, in this hot air, and with this immense horizon, is very interesting. Showers in this tentative time are local. In our journey we saw showers far off, we experienced a dash for ten minutes, but it was local, covering not more than a mile or two square. We have in sight a vast canopy of blue sky, of forming and dispersing clouds. It is difficult for them to drop their moisture in the rising columns of hot air. The result at times was a very curious spectacle—rain in the sky that did not reach the earth. Perhaps some cold current high above us would condense the moisture, which would begin to fall in long trailing sweeps, blown like fine folds of muslin, or like sheets of dissolving sugar, and then the hot air of the earth would dissipate it, and the showers would be absorbed in the upper regions. The heat was sometimes intense, but at intervals a refreshing wind would blow, the air being as fickle as the rain; and now and then we would see a slender column of dust, a thousand or two feet high, marching across the desert, apparently not more than two feet in diameter, and wavering like the threads of moisture that tried in vain to reach the earth as rain. Of life there was not much to be seen in our desert route. In the first day we encountered no habitation except the ranch-house mentioned, and saw no human being; and the second day none except the solitary occupant of the dried well at Red Horse, and two or three Indians on the hunt. A few squirrels were seen, and a rabbit now and then, and occasionally a bird. The general impression was that of a deserted land. But antelope abound in the timber regions, and we saw several of these graceful creatures quite near us. Excellent antelope steaks, bought of the wandering Indian hunters, added something to our "canned" supplies. One day as we lunched, without water, on the cedar slope of a lovely grass interval, we saw coming towards us over the swells of the prairie a figure of a man on a horse. It rode to us straight as the crow flies. The Indian pony stopped not two feet from where our group sat, and the rider, who was an Oualapai chief, clad in sacking, with the print of the brand of flour or salt on his back, dismounted with his Winchester rifle, and stood silently looking at us without a word of salutation. He stood there, impassive, until we offered him something to eat. Having eaten all we gave him, he opened his mouth and said, "Smoke 'em?" Having procured from the other wagon a pipe of tobacco and a pull at the driver's canteen, he returned to us all smiles. His only baggage was the skull of an antelope, with the horns, hung at his saddle. Into this he put the bread and meat which we gave him, mounted the wretched pony, and without a word rode straight away. At a little distance he halted, dismounted, and motioned towards the edge of the timber, where he had spied an antelope. But the game eluded him, and he mounted again and rode off across the desert—a strange figure. His tribe lives in the cañon some fifty miles west, and was at present encamped, for the purpose of hunting, in the pine woods not far from the point we were aiming at.

The way seemed long. With the heat and dust and slow progress, it was exceedingly wearisome. Our modern nerves are not attuned to the slow crawling of a prairie-wagon. There had been growing for some time in the coach a feeling that the journey did not pay; that, in fact, no mere scenery could compensate for the fatigue of the trip. The imagination did not rise to it. "It will have to be a very big cañon," said the duchess.

Late in the afternoon we entered an open pine forest, passed through a meadow where the Indians had set their camp by a shallow pond, and drove along a ridge, in the cool shades, for three or four miles. Suddenly, on the edge of a descent, we who were on the box saw through the tree-tops a vision that stopped the pulse for a second, and filled us with excitement. It was only a glimpse, far off and apparently lifted up—red towers, purple cliffs, wide-spread apart, hints of color and splendor; on the right distance, mansions, gold and white and carmine (so the light made them), architectural habitations in the sky it must be, and suggestions of others far off in the middle distance—a substantial aerial city, or the ruins of one, such as the prophet saw in a vision. It was only a glimpse. Our hearts were in our mouths. We had a vague impression of something wonderful, fearful—some incomparable splendor that was not earthly. Were we drawing near the "City?" and should we have yet a more perfect view thereof? Was it Jerusalem or some Hindoo temples there in the sky? "It was builded of pearls and precious stones, also the streets were paved with gold; so that by reason of the natural glory of the city, and the reflection of the sunbeams upon it, Christian with desire fell sick." It was a momentary vision of vast amphitheatre of splendor, mostly hidden by the trees and the edge of the plateau.

We descended into a hollow. There was the well, a log-cabin, a tent or two under the pine-trees. We dismounted with impatient haste. The sun was low in the horizon, and had long withdrawn from this grassy dell. Tired as we were, we could not wait. It was only to ascend the little steep, stony slope—300 yards—and we should see! Our party were straggling up the hill: two or three had reached the edge. I looked up. The duchess threw up her arms and screamed. We were not fifteen paces behind, but we saw nothing. We took the few steps, and the whole magnificence broke upon us. No one could be prepared for it. The scene is one to strike dumb with awe, or to unstring the nerves; one might stand in silent astonishment, another would burst into tears.

There are some experiences that cannot be repeated—one's first view of Rome, one's first view of Jerusalem. But these emotions are produced by association, by the sudden standing face to face with the scenes most wrought into our whole life and education by tradition and religion. This was without association, as it was without parallel. It was a shock so novel that the mind, dazed, quite failed to comprehend it. All that we could grasp was a vast confusion of amphitheatres and strange architectural forms resplendent with color. The vastness of the view amazed us quite as much as its transcendent beauty.

We had expected a cañon—two lines of perpendicular walls 6000 feet high, with the ribbon of a river at the bottom; but the reader may dismiss all his notions of a cañon, indeed of any sort of mountain or gorge scenery with which he is familiar. We had come into a new world. What we saw was not a cañon, or a chasm, or a gorge, but a vast area which is a break in the plateau. From where we stood it was twelve miles across to the opposite walls—a level line of mesa on the Utah side. We looked up and down for twenty to thirty miles. This great space is filled with gigantic architectural constructions, with amphitheatres, gorges, precipices, walls of masonry, fortresses terraced up to the level of the eye, temples mountain size, all brilliant with horizontal lines of color—streaks of solid hues a few feet in width, streaks a thousand feet in width—yellows, mingled white and gray, orange, dull red, brown, blue, carmine, green, all blending in the sunlight into one transcendent suffusion of splendor. Afar off we saw the river in two places, a mere thread, as motionless and smooth as a strip of mirror, only we knew it was a turbid, boiling torrent, 6000 feet below us. Directly opposite the overhanging ledge on which we stood was a mountain, the sloping base of which was ashy gray and bluish; it rose in a series of terraces to a thousand-feet wall of dark red sandstone, receding upward, with ranges of columns and many fantastic sculptures, to a finial row of gigantic opera-glasses 6000 feet above the river. The great San Francisco Mountain, with its snowy crater, which we had passed on the way, might have been set down in the place of this one, and it would have been only one in a multitude of such forms that met the eye whichever way we looked. Indeed, all the vast mountains in this region might be hidden in this cañon.

Wandering a little away from the group and out of sight, and turning suddenly to the scene from another point of view, I experienced for a moment an indescribable terror of nature, a confusion of mind, a fear to be alone in such a presence. With all this grotesqueness and majesty of form and radiance of color, creation seemed in a whirl. With our education in scenery of a totally different kind, I suppose it would need long acquaintance with this to familiarize one with it to the extent of perfect mental comprehension.

The vast abyss has an atmosphere of its own, one always changing and producing new effects, an atmosphere and shadows and tones of its own—golden, rosy, gray, brilliant, and sombre, and playing a thousand fantastic tricks to the vision. The rich and wonderful color effects, says Captain Dutton, "are due to the inherent colors of the rocks, modified by the atmosphere. Like any other great series of strata in the plateau province, the carboniferous has its own range of colors, which might serve to distinguish it; even if we had no other criterion. The summit strata are pale gray, with a faint yellowish cast. Beneath them the crossbedded sandstone appears, showing a mottled surface of pale pinkish hue. Underneath this member are nearly 1000 feet of the lower Aubrey sandstones, displaying an intensely brilliant red, which is somewhat marked by the talus shot down from the gray cherty limestone at the summit. Beneath the lower Aubrey is the face of the Red Wall limestone, from 2000 to 3000 feet high. It has a strong red tone, but a very peculiar one. Most of the red strata of the West have the brownish or vermilion tones, but these are rather purplish red, as if the pigment had been treated to a dash of blue. It is not quite certain that this may not arise in part from the intervention of the blue haze, and probably it is rendered more conspicuous by this cause; but, on the whole, the purplish cast seems to be inherent. This is the dominant color of the cañon, for the expanse of the rock surface displayed is more than half in the Red Wall Group."

I was continually likening this to a vast city rather than a landscape, but it was a city of no man's creation nor of any man's conception. In the visions which inspired or crazy painters have had of the New Jerusalem, of Babylon the Great, of a heaven in the atmosphere, with endless perspective of towers and steeps that hang in the twilight sky, the imagination has tried to reach this reality. But here are effects beyond the artist, forms the architect has not hinted at; and yet everything reminds us of man's work. And the explorers have tried by the use of Oriental nomenclature to bring it within our comprehension, the East being the land of the imagination. There is the Hindoo Amphitheatre, the Bright Angel Amphitheatre, the Ottoman Amphitheatre, Shiva's Temple, Vishnu's Temple, Vulcan's Throne. And here, indeed, is the idea of the pagoda architecture, of the terrace architecture, of the bizarre constructions which rise with projecting buttresses, rows of pillars, recesses, battlements, esplanades, and low walls, hanging gardens, and truncated pinnacles. It is a city, but a city of the imagination. In many pages I could tell what I saw in one day's lounging for a mile or so along the edge of the precipice. The view changed at every step, and was never half an hour the same in one place. Nor did it need much fancy to create illusions or pictures of unearthly beauty. There was a castle, terraced up with columns, plain enough, and below it a parade-ground; at any moment the knights in armor and with banners might emerge from the red gates and deploy there, while the ladies looked down from the balconies. But there were many castles and fortresses and barracks and noble mansions. And the rich sculpture in this brilliant color! In time I began to see queer details: a Richardson house, with low portals and round arches, surmounted by a Nuremberg gable; perfect panels, 600 feet high, for the setting of pictures; a train of cars partly derailed at the door of a long, low warehouse, with a garden in front of it. There was no end to such devices.

It was long before I could comprehend the vastness of the view, see the enormous chasms and rents and seams, and the many architectural ranges separated by great gulfs, between me and the wall of the mesa twelve miles distant. Away to the northeast was the blue Navajo Mountain, the lone peak in the horizon; but on the southern side of it lay a desert level, which in the afternoon light took on the exact appearance of a blue lake; its edge this side was a wall thousands of feet high, many miles in length, and straightly horizontal; over this seemed to fall water. I could see the foam of it at the foot of the cliff; and below that was a lake of shimmering silver, in which the giant precipice and the fall and their color were mirrored. Of course there was no silver lake, and the reflection that simulated it was only the sun on the lower part of the immense wall.

Someone said that all that was needed to perfect this scene was a Niagara Falls. I thought what figure a fall 150 feet high and 3000 long would make in this arena. It would need a spy-glass to discover it. An adequate Niagara here should be at least three miles in breadth, and fall 2000 feet over one of these walls. And the Yosemite—ah! the lovely Yosemite! Dumped down into this wilderness of gorges and mountains, it would take a guide who knew of its existence a long time to find it.

The process of creation is here laid bare through the geologic periods. The strata of rock, deposited or upheaved, preserve their horizontal and parallel courses. If we imagine a river flowing on a plain, it would wear for itself a deeper and deeper channel. The walls of this channel would recede irregularly, by weathering and by the coming in of other streams. The channel would go on deepening, and the outer walls would again recede. If the rocks were of different material and degrees of hardness, the forms would be carved in the fantastic and architectural manner we find them here. The Colorado flows through the tortuous inner chasm, and where we see it, it is 6000 feet below the surface where we stand, and below the towers of the terraced forms nearer it. The splendid views of the cañon at this point given in Captain Dutton's report are from Point Sublime, on the north side. There seems to have been no way of reaching the river from that point. From the south side the descent, though wearisome, is feasible. It reverses mountaineering to descend 6000 feet for a view, and there is a certain pleasure in standing on a mountain summit without the trouble of climbing it. Hance, the guide, who has charge of the well, has made a path to the bottom. The route is seven miles long. Halfway down he has a house by a spring. At the bottom, somewhere in those depths, is a sort of farm, grass capable of sustaining horses and cattle, and ground where fruit-trees can grow. Horses are actually living there, and parties descend there with tents, and camp for days at a time. It is a world of its own. Some of the photographic views presented here, all inadequate, are taken from points on Hance's trail. But no camera or pen can convey an adequate conception of what Captain Dutton happily calls a great innovation in the modern ideas of scenery. To the eye educated to any other, it may be shocking, grotesque, incomprehensible; but "those who have long and carefully studied the Grand Cañon of the Colorado do not hesitate for a moment to pronounce it by far the most sublime of all earthly spectacles."

I have space only to refer to the geologic history in Captain Dutton's report of 1882, of which there should be a popular edition. The waters of the Atlantic once overflowed this region, and were separated from the Pacific, if at all, only by a ridge. The story is of long eras of deposits, of removal, of upheaval, and of volcanic action. It is estimated that in one period the thickness of strata removed and transported away was 10,000 feet. Long after the Colorado began its work of corrosion there was a mighty upheaval. The reader will find the story of the making of the Grand Cañon more fascinating than any romance.

Without knowing this story the impression that one has in looking on this scene is that of immense antiquity, hardly anywhere else on earth so overwhelming as here. It has been here in all its lonely grandeur and transcendent beauty, exactly as it is, for what to us is an eternity, unknown, unseen by human eye. To the recent Indian, who roved along its brink or descended to its recesses, it was not strange, because he had known no other than the plateau scenery. It is only within a quarter of a century that the Grand Cañon has been known to the civilized world. It is scarcely known now. It is a world largely unexplored. Those who best know it are most sensitive to its awe and splendor. It is never twice the same, for, as I said, it has an atmosphere of its own. I was told by Hance that he once saw a thunderstorm in it. He described the chaos of clouds in the pit, the roar of the tempest, the reverberations of thunder, the inconceivable splendor of the rainbows mingled with the colors of the towers and terraces. It was as if the world were breaking up. He fled away to his hut in terror.

The day is near when this scenery must be made accessible. A railway can easily be built from Flagstaff. The projected road from Utah, crossing the Colorado at Lee's Ferry, would come within twenty miles of the Grand Cañon, and a branch to it could be built. The region is arid, and in the "sight-seeing" part of the year the few surface wells and springs are likely to go dry. The greatest difficulty would be in procuring water for railway service or for such houses of entertainment as are necessary. It could, no doubt, be piped from the San Francisco Mountain. At any rate, ingenuity will overcome the difficulties, and travellers from the wide world will flock thither, for there is revealed the long-kept secret, the unique achievement of nature.

# CHAPTER THREE

# 1899
# The Grand Canyon
# of the Colorado

## Harriet Monroe

Fashions in journalism have changed since the 1890s, so that what then seemed to be rhapsody now is gushing sentimentality. Harriet Monroe, a successful Chicago poet-essayist, portrayed the canyon with all proper Victorian womanliness, as if her frail female senses were not strong enough to cope with such a wonder. Her overwrought prose may entertain us for a different reason than she intended, but her point is still valid. Among the many responses today's visitors have to the Grand Canyon, humility and a feeling of sensory inadequacy are two of the most common.

> The earth grew bold with longing
> And called the high gods down:
> Yea, though ye dwell in heaven and hell,
> I challenge their renown.
> Abodes as fair I build ye
> As heaven's rich courts of pearl,
> And chasms dire where floods like fire
> Ravage and roar and whirl.
> Come, for my soul is weary

Of time and death and change;
Eternity doth summon me—
With mightier worlds I range.
Come, for my vision's glory
Awaits your songs and wings;
Here on my breast I bid ye rest
From starry wanderings.

THE sun-browned miner who sat opposite me in the dusty stage talked of our goal to shorten the long hours of the journey, and of the travelers who had preceded us over that lonely trail to the edge of the Grand Cañon of the Colorado River. "Yes, I have been in and out of the cañon for twenty years," he said, "and I haven't begun to understand it yet. The Lord knows, perhaps, why he gave it to us; I never felt big enough to ask." And he told the story of a young English preacher whom he once picked up near the end of the road; who, too poor to pay stage fares, was walking to the cañon; who, after two days and nights in the thirsty wastes, his canteen empty and only a few biscuits left in his pouch, was trudging bravely on, with blistered feet and aching body, because he "must see" the mighty miracle beyond.

We were out in the open endless desert, the sunburned desolate waste. Our four horses kicked up the dust of the road, and the wind whirled it into our faces and sifted it through our clothes. We had passed the halfway house, where, finding the shanty too hot, we had unpacked and eaten our luncheons out in the sun and wind. It was just at the weary moment of the long, hot drive when the starting place seemed lost in the past, and the goal still far ahead; but the thought of the preacher's ardor made us ashamed to be tired, gave us back the beauty of the day. All the morning we had driven through forests of tall pines and bare white aspens, watching the changing curves of San Francisco Mountain, whose lofty head rose streaked with white against the blue; until at last, as we rounded its foothills, the desert lay below us like a sea, and we descended to the magic shore and took passage over the billows of silver and amethyst that foamed and waved beyond and afar. Lines of opalescent light grew into rocky mesas rising steep and formidable out of the barren plain. Silvery vistas widened into deserts so barren that even sagebrush and dwarfish cactus choked there; and the only signs of life, paradoxically, were the chalk-white skeletons of animals that lay collapsing into dust beside the road. All day long we were alone with the world's immensity—no human face or voice breaking the wastes of forest and plain, except when our tired horses thrice gave way to fresh ones, and their keepers came out from little shacks to unbuckle the harness and hear the news.

The immense and endless desolation seemed to efface us from the earth. What right had we there, on those lofty lands which never since the beginning of time had offered sustenance to man? Since first the vast plain with its mighty weight of mountains arose far out of the waters, no kindly rill or fountain had broken the silence and invited life. What hidden wells would feed the prairie dogs, what rains would slake the large thirst of the pines, while now for months the aching land must parch and burn under a cloudless sky? It was May, and yet the summer had begun in these high places of the earth, and the last flecks of snow were fading from the peaks. Following slowly the gentle grades of the road, we tried to appreciate the altitude. Was it possible that these long levels lay a mile and a half above the ocean; that this barren slope, where the wind blew keen, was only a thousand feet nearer earth than the crest of the Dent du Midi, whose notched and snowy peak dominates Lac Lèman? No wonder the waters leave the great plateau to the sun, and hurl themselves against moun-tainous barriers, and carve out gorges and cañons in their wild eagerness to find the sea!

At last we reach the third relay station, and take on six horses instead of four, for the final pull uphill. We alight, and run up and down the shaggy little slope, and free our bodies from the long strain. We reflect that as we are traveling now, even in this primi-tive slavery to beasts of burden, so for many centuries our fathers had traversed the earth, knowing no swifter way. All day for seventy-five miles—what a tyrannous abuse of time! And yet through ages and ages the lords of the earth had been so deaf to its voices that not one secret of nature's power had escaped to help them conquer her. We had left the nineteenth century behind; we were exploring the wilderness with the pioneers. We were unaware of the road, of the goal; we were pushing out into the un-known, buffeted by its denials, threatened by its wars, lured by its mysteries. The desert lay behind us now; once more the quiet forest for miles on miles. So still and sweet and sylvan were its smooth brown slopes; the tallest pines whose vision overtopped their neighbors were all unsuspicious of nature's appalling and magnificent intention. And we, we could not believe that the forest would not go on forever, even when vistas of purple began to open through the trees, even when the log-cabin hotel welcomed us to our goal.

It was like sudden death—our passing round the corner to the other side of that primitive inn; for in a moment we stood at the end of the world, at the brink of the kingdoms of peace and pain. The gorgeous purples of sunset fell into darkness and rose into light over mansions colossal beyond the needs of our puny unwinged race. Terrific abysses yawned and darkened; magical heights glowed with iridescent fire. The earth lay stricken to the heart, her masks and draperies torn away, confessing her eternal passion to the absolving sun. And even as we watched and hearkened, the pitiful night lent deep shadows to cover her majesty and hide its awful secrets from the curious stars.

In the morning, when I went out to verify the vision, to compass earth's revelation of her soul, the sun fell to the very heart of the mystery, even from the depths rose a thrill of joy. It was morning; I had slept and eaten; the fatigue and dust of the long journey no longer oppressed me; my courage rose to meet the greatness of the world. The benevolent landlady told of a trail which led to Point Lookout, a mile and a half away, beneath whose cliffs the old deserted inn lay in a hollow. I set out with two companions of the stage, who were armed with cameras and possessed of modern ideas. They pleaded for improvements: built a railroad from Flagstaff to the rim, a summer hotel on one of those frowning cliffs; yes, even a funicular railway down to the hidden river, and pumping works which should entice its waters up the steep slope to the thirsty beasts and travelers whose drink must now be hauled from the halfway house, forty miles away. But I rose up and defended the wilderness; rejoiced in the dusty stage ride, in the rough cabin that rose so fitly from the clearing, in the vast unviolated solitudes—in all these proofs that one of the glories of earth was still undesecrated by the chatter of facile tourists; that here we must still propitiate nature with sacrifices, pay her with toil, prove the temper of our souls before assailing her immensities. And when my companions accused me of selfishness, opened the hidden wonder to all the world, and made it the common property of literature and art, the theme of all men's praise, even like Mont Blanc and the Colosseum and Niagara, my tongue had no words of defense to utter, but my heart rejoiced the more that I had arrived before all these.

We wandered along the quietest sylvan path, which led us up and down little ravines and dales, always under the shade of tall pines, always over the brown carpet of their needles. Now and then a sudden chasm would lift a corner of the veil, and we would wonder how we dared go on. Yet on and on we went—a mile and a half, two miles, three—and still no deserted cabin under slanting cliffs. My companions recalled the landlady's words, were sure that we had missed the road, and resolved to go back and find it; so I urged them to the search, and promised to rest and follow.

Tollgate at the head of Bright Angel Trail, which operated from 1903 to 1924. *Courtesy of the Grand Canyon Natural History Association*

But when I had rested the trail allured me; surely it was too clear to lead me wrong. I would explore it yet a little. I walked on—five minutes, ten—and there below me lay the hollow and the cabin. I passed it, the little silent lodge, with rough-hewn seats under the broad eaves of its porch, its doors hospitably unlatched, its rooms still rudely furnished; but all dusty, voiceless, forsaken. I climbed the steep slope to the rocks, crawled half prostrate to the barest and highest, and lay there on the edge of the void, the only living thing in some unvisited world.

For surely it was not our world, this stupendous, adorable vision. Not for human needs was it fashioned, but for the abode of gods. It made a coward of me; I shrank and shut my eyes, and felt crushed and beaten under the intolerable burden of the flesh. For humanity intruded here; in these warm and glowing purple spaces disembodied spirits must range and soar, souls purged and purified and infinitely daring. I felt keenly sure of mighty presences among the edifices vast in scope and perfect in design that rose from the first foundations of the earth to the lofty level of my jagged rock. Prophets and poets had wandered here before they were born to tell their mighty tales—Isaiah and Aeschylus and Dante, the giants who dared the utmost. Here at last the souls of great architects must find their dreams fulfilled; must recognize the primal inspiration which, after long ages, had achieved Assyrian palaces, the temples and pyramids of Egypt, the fortresses and towered cathedrals of mediaeval Europe. For the inscrutable Prince of builders had reared these imperishable monuments, evenly terraced upward from the remote abyss; had so cunningly planned them that mortal foot could never climb and enter, to disturb the everlasting hush. Of all richest elements they were fashioned—jasper and chalcedony, topaz, beryl, and amethyst, firehearted opal and pearl; for they caught and held the most delicate colors of a dream, and flashed full recognition to the sun. Never on earth could such glory be unveiled—not on level spaces of sea, not on the cold bare peaks of mountains. This was not earth; for was not heaven itself across there, rising above yonder alabaster marge in opalescent ranks for the principalities and powers? This was not earth—I intruded here. Everywhere the proof of my unfitness abased and dazed my will: this vast unviolated silence, as void of life and death as some newborn world; this mystery of omnipotence revealed, laid bare, but incomprehensible to my weak imagining; this inaccessible remoteness of depths and heights, from the sinuous river which showed afar one or two tawny crescents curving out of impenetrable shadows, to the mighty temple of Vishnu which guilded its vast tower loftily in the sun. Not for me, not for human souls, not for any form of earthly life, was the secret of this unveiling. Who that breathed could compass it?

The strain of existence became too tense against these infinities of beauty and terror. My narrow ledge of rock was a prison. I fought against the desperate temptation to fling myself down into that soft abyss, and thus redeem the affront which the eager beating of my heart offered to its inviolable solitude. Death itself would not be too rash an apology for my invasion—death in those happy spaces, pillowed on purple immensities of air. So keen was the impulse, so slight at that moment became the fleshly tie, that I might almost have yielded but for a sudden word in my ear—the trill of an oriole from the pine close above me. The brave little song was a message personal and intimate, a miracle of sympathy or prophecy. And I cast myself on that tiny speck of life as on the heart of a friend—a friend who would save me from intolerable loneliness, from utter extinction and despair. He seemed to welcome me to the infinite; to bid me go forth and range therein, and know the lords of heaven and earth who there had drunk the deep waters and taken the measure of their souls. I made him the confidant of my unworthiness; asked him for the secret, since, being winged, he was at home even here. He gave me healing and solace; restored me to the gentle amenities of our little world; enabled me to retreat through the woods, as I came, instead of taking the swift dramatic road to liberty.

I do not know how one could live long on the rim of that abyss of glory, on the brink of sensations too violent for the heart of man. I looked with wonder at the guides and innkeepers, the miners and carriers, for whom the utmost magnificence of earth is the mere background of daily living. Does it crush or inspire? Do they cease to feel it, or does it become so close a need that all earth's fields and brooks and hills are afterward a petty prison for hearts heavy with longing? When they go down to the black Inferno where that awful river still cuts its way through the first primeval shapeless rocks, where the midday darkness reveals night's stars in a cleft of sky, while the brown torrent roars and laughs at its frowning walls—when they, mere men of the upper air, descend to that nether world, do they recognize the spirits of darkness who shout and strain and labor there? And when they emerge, and step by step ascend the shining cliffs, do they feel like Dante when he was led by his celestial love to paradise?

The days of my wanderings along the edge of the chasm were too few to reconcile my littleness with its immensity. To the end it effaced me. I found comfort in the forests, whose gentle and comprehensible beauty restored me to our human life. It was only the high priest who could enter the Holy of Holies, and he only once a year; so here, in nature's innermost sanctuary, man must be of the elect, must purify his soul with fasting and prayer and clothe it in fine raiment, if he would worthily tread the sacred ground. It is not for nothing that the secret is hidden in the wilderness, and that the innermost depths of it are inaccessible to our wingless race. At this point one or two breakneck trails lead down to the Styx-like river, but he who descends to the dark waters must return by the same road; he may not follow the torrent through the bowels of the earth except to be its sport or prey. Even though he embarks upon that fearsome journey, and even though, like Major Powell and his handful of adventurers, he escapes death by a thousand miracles, yet he may not emerge from the depths of hell through all the days and nights of the journey; he may not set foot on the purple slopes and climb to the pearly mansions—nay, nor even behold them, overshadowed as he is by frowning walls that seem to cut the sky. For a few miles along the rim and down a trail or two to the abyss, human feet and human eyes may risk body and soul for an exceeding great reward; but for a hundred miles beyond, both to right and left, the mystery is still inviolate. He who attempts it dies of thirst in the desert, or of violence in the chasm.

Tragic stories are told of men who have lost their lives in the search for precious metals which may lie hidden or uncovered here. The great primeval flood cut its broad V through all the strata of rock, with all their veins of metallic ore, down to the earliest shapeless mass, leaving in its wake the terraced temples and towers which seem to have been planned by some architect of divinest genius to guard their treasures inviolate till the end of time. And the river, rising far to the north among mountains rich in mineral, has been washing away the sand for ages, and depositing its gold and silver and lead in the still crevices of the impenetrable chasm. Here the earth laughs at her human master, and bids him find her wealth if he dare, and bear it away if he can. A young Californian who accepted the challenge, and set forth upon the turgid water to sift its sands for gold, never emerged with his hapless men to tell the story of his search. Only near the brink of the cleft are a few miners burrowing for copper, and sending their ore up to the rim on the backs of hardy burros; as who should prick the mountain with a pin, or measure the ocean with a cup.

As I grew familiar with the vision, I could not quite explain its stupendous quality. From mountain tops one looks across greater distances, and sees range after range lifting snowy peaks into the blue. The ocean reaches out into boundless space, and the ebb and flow of its waters have the beauty of rhythmic motion and exquisitely varied color. And in the rush of mighty cataracts are power and splendor and majestic peace. Yet for grandeur appalling and unearthly, for ineffable, impossible beauty, the cañon transcends all these. It is as though to the glory of nature were added the glory of art; as though, to achieve her utmost, the proud young world had commanded architecture to build for her and color to grace the building. The irregular masses of mountains, cast up out of the molten earth in some primeval war of elements, bear no relation to these prodigious symmetrical edifices, mounted on abysmal terraces and grouped into spacious harmonies which give form to one's dreams of heaven. The sweetness of green does not last forever, but these mightily varied purples are eternal. All that grows and moves must perish, while these silent immensities endure. Lovely and majestic beyond the cunning of human thought, the mighty monuments rise to the sun as lightly as clouds that pass. And forever glorious and forever immutable, they must rebuke man's pride with the vision of ultimate beauty, and fulfill earth's dream of rest after her work is done.

# CHAPTER FOUR

# 1902

# To the Grand Canyon on an Automobile

Winfield Hogaboom

Today many people look upon automobiles as a mixed blessing in national parks. Some of the youngest parks have very few public roads, and others are instituting mass transit systems in order to keep car traffic to a minimum and prevent disruptions of fragile natural areas. Even when cars were new they had their detractors, but few people in 1902 realized how important the car would eventually become in American life.

A year after the trip described in this chapter, a Vermont physician completed the first transcontinental automobile trip. It took him nine weeks and cost $8,000.[1] Pioneering automobilists like these were not followed by a flood of touring cars, but they truly were opening up a new frontier. By 1926, as many people reached the Grand Canyon by car as by rail, and today practically all visitors to national parks arrive by car.

In spite of what seems to have been an arduous and unpleasant first attempt, the driver in Hogaboom's account was quite enthusiastic about the prospects of regular automobile traffic to the canyon. In fact, the newspaper in Flagstaff was quick to announce that "it is quite probable that a company will be organized here to establish an automobile line between this place and the Grand Canyon."[2] As events would show, no such company was necessary.

THE automobile has gone the limit, as the saying is. The Grand Canyon of the Colorado, that fearful gash in the fair face of Nature, that has gone far towards making Arizona world-famous; the Painted desert, that weird and awful stretch of arid plain, where it seems as if two-thirds of all outdoors might be dropped and lost; the vast Coconino forest, with its miles upon miles of somber shadow and unearthly silence; and the stolid Walpi and Navajo Indians, sad remnants of another day and a once mighty race, all have been visited by the "fire wagon" that runs itself.

The automobile has had its picture taken standing upon the rim of the mighty chasm, and it actually succeeded in looking pleasant while it hung on the brink of the biggest hole in the ground in all the world. Fifteen miles or more away, as the airship flies, was the opposite rim of the canyon, and a sheer mile and a half downward was what seems from above to be a deep gorge, through which the Colorado river winds like a silver thread. It is a deep gorge, as anyone who goes down to it will discover. I'd hate to fall into it.

But before it came to this outlandish place to have its picture taken with the world's greatest natural wonder for a background, the automobile went through some strange, eventful experiences that no other automobile ever had. It scooted out of a little Arizona city one sunshiny afternoon, while the people were waving handkerchiefs and hats, and shouting good luck and good-by. Just outside the town it climbed a small hill, stood upon the summit for an instant silhouetted against the clear, blue sky, and gliding down the other slope, was lost to view.

The automobile then found itself in a beautiful little valley, and it just had time to find out that this was a peculiar sort of a valley, where nobody lives, and where the only house is an old log fort, built when the Indians were troublesome, when it discovered also that it had crossed the valley and was about to plunge into a vast forest of giant pines. But it never faltered. It was prepared for forests or anything else.

The automobile traveled through the forest long after the sun had gone down; long after the light had faded, and the slanting shadows of the pines were blended, and the stars shone brightly in the blue vault above the dark outlines of the trees. But something got the matter with it after awhile, and it didn't want to travel any more. So it stayed all night in the forest.

Next day it went on. It emerged from the forest after awhile and struck out across the open plain, past a wonderful mountain of red lava cinders, past deserted Indian villages, past bands of graceful antelope and deer and wild horses. When darkness came again the automobile was in the midst of another forest, a forest of cedars, and there it stayed for another night.

Next day the automobile was completely tired out—at least it refused to go any more—and was left right there until it was rested up. It arrived at the Grand Canyon on the afternoon of the day after the next. And great was the wonderment of the few people who live on the rim of the big hole when they saw it there.

The idea of running this particular automobile from Flagstaff, on the line of the Santa Fe road, to the Grand Canyon, sixty-five miles away, was incubated in the fertile brain of Oliver Lippincott of Los Angeles. For years stages had been running between Flagstaff and the canyon until the railroad up and built a line that branches from the main line of the Santa Fe system at Williams, and runs to Bright Angel hotel, on the rim. The stages could not compete with the railroad. Somebody suggested that automobiles might do so, and last summer a man took a little gasoline automobile out to Flagstaff and announced that he was ready to try it.

The people of Flagstaff hailed the advent of the automobile with delight and did everything they could to help. But the man never made the attempt. He laid plans for the trip on many different occasions, and just at the last moment he always discovered that something was wrong with his machine, and postponed the trial. At length he loaded his little auto onto a flatcar and went away with it, and the people of Flagstaff never saw him any more.

So when Mr. Lippincott came with his big automobile they had very little to say—and what little they did say wasn't encouraging. They were willing to help, of course. The people of Flagstaff are peculiar that way. They will help anybody or anything. There is something in the atmosphere there—just the right altitude, or something—that makes the Flagstaff people kindly. But they all had a little quiet snicker or two, on the side.

Mr. Lippincott went right along, however, and didn't look for any cheering words. With him were T. N. Chapman, a newspaperman, formerly of New York, but now living in Los Angeles, and myself. At Flagstaff Mr. Lippincott engaged the services of Al Doyle as guide. Mr. Doyle is unique. He is a little man, but there is something about him that suggests strength and grit and resource. He is an oldtimer in those parts: a plainsman, a mountaineer, and a typical westerner.

## An Idea of the Country

So the crew was composed of four people. The automobile was one that the Toledo Automobile company had built and sent out to Flagstaff on purpose to make this trip. It is a steam machine, with a ten horse-power engine of the type used by the government on its torpedo boats, with a coil boiler and using gasoline for fuel. On the main machine was a seat for two persons, the driver and the guide, while the other two members of the crew were seated on a trailer, carrying, besides the two persons, something like a thousand pounds of baggage and also water and gasoline for the boiler.

I want to give, as near as I can recall them, Doyle's directions in regard to the route to be followed, as he gave them to Mr. Lippincott the night we arrived in Flagstaff. They afford some idea of the country between Flagstaff and the canyon, and also an idea of Doyle.

"After you get over the rise just outside of town," Doyle began, "you strike down through Fort valley, and bear off to the west, so's not to get too far up onto the foothills while you're getting by the peaks. You hit the new trail there and cross the big wash, and pretty soon you get into the trees. Dry Tanks is about four miles east of you, and after you pass Sheep Ranch and come up out of the gulch you turn east again, till you get to the clearing. The trail goes right straight into Moodyspan's cabin, and then you strike into the trees again until you get to Red mountain. Then you take off across the desert and follow the old sheep trail till you strike the mouth of Red Horse canyon. Just left of the Dog Knobs you come into the cedars again, and keep right through on the old trail till you get to Skinner's cabin. Then you have to bear off down the gulch to the right and follow the gulch. After you strike the old trail back of Red buttes you follow it right in to Berry's."

Immediately upon getting this accurate description of our route, Lippincott decided to take Doyle along with us, for luck.

## The Start From Flagstaff

We started at 2:10 o'clock on Saturday afternoon, January 4. The people of Flagstaff were out en masse to see us off, and it was amusing to listen to the conjectures they made in regard to our trip. They gave us anywhere from six days to six years in which to make the journey.

The machine worked splendidly—until we were out of sight of the assembled populace. We were thankful for that. But before we had covered the first ten miles it got to acting up, and our chauf-

Hogaboom and company in their automobile at the rim of the Grand Canyon. *Courtesy of the Grand Canyon Natural History Association*

feur—whatever the deuce that means—said that the trailer was bearing down too hard on the "hind" axle. So we got off, unloaded the outfit from the trailer and fixed it in such a manner that it couldn't bear down too hard. And when we got through with that job and got the baggage and ourselves aboard again darkness had fallen upon us.

Ahead of us loomed the great Coconino forest. I cannot describe it. There is no underbrush, no small trees, just the giant pines sticking straight up to the sky out of the ground. And the ground is smooth and hard, and carpeted with pine needles. It was dark in the forest, for the tops of these trees are big and dense and only here and there the light from above trickles through. But we could not stop for darkness. We sped on and on. Fantastic forms appeared around us—forms of bears and tigers and elephants and alligators and the prehistoric animals that thrive in the Smithsonian Institution and the Sunday papers; terrible monsters that we feared might bite us.

But we went by so fast that they only stared at us and disappeared, and Chauffeur Lippincott said that if we didn't run against any of those fallen trees or charred stumps we would get through all right. We hoped so, but it was hard to believe.

It was a cold night—cold for us thin-blooded Californians at least—and we suffered some.

## Cowboy Hospitality

After about two hours of this thing Doyle, the guide, suggested a halt. Ahead of us, he said, there was a particularly bad stretch of country, and it would be better to go over it in daylight. This was a disappointment to all of us, for we had promised the people of Flagstaff that we would make the whole journey to the Grand canyon that night, and what is more we had really expected to do so. And to prove to them that we really expected to, we had made no provision for a stop at night, and had no eatables along. All that we had in the way of the necessities of life was plenty of tobacco and matches.

But it is one thing to say that you are going to run an automobile through a wild, uninhabited country at night, when the moon is off duty, and another thing to do it. After a short consultation we decided to stop.

Right here our first streak of good luck came to us. Doyle reckoned that Moodyspan's cabin was somewhere in the vicinity. We came to where a side trail led away from ours, up the side of a mountain, and he reckoned that trail led right to the cabin. We had to leave the automobile, however, for it wasn't made to climb mountains like that one, and to scramble on foot up the winding trail. It seemed a long way, but after awhile we saw something that made us glad—a light apparently shining right out of the side of the big, dark mountain. Doyle reckoned somebody must be staying at the cabin. And so there was. Three cowboys were staying there. Ordinarily the cabin is deserted. We had expected nothing more than shelter. We found cowboys and cowboy hospitality. They had recently killed a 2-year-old steer, and they also had flour, baking powder, coffee, honey and ice. The ice was brought from the mountain of ice in the wonderful bottomless pits that are found seven miles distant from where we then were. What more could a person reasonably ask for—juicy steaks, biscuits, honey and coffee; a supper fit for hungry automobilists.

We bunked with the cowboys on the floor of the cabin, and were up long before daylight and had a breakfast of coffee, honey, biscuits and steaks. We were back where we had left the automobile just as the first gray dawn was breaking. The night had been very cold. The automobile was frozen stiff. A long time was required to thaw it out, and a valuable portion of our supply of fuel was used. But we thought little of that, for were we not going to be at Berry's place, on the rim, by 2 o'clock of that day at the very latest?

Alas! for our reckoning, and alas! for us. Alas! several times for us. We got to Berry's two days later, and not a particle of food did we have to eat, nor a drop of water to drink, until we got there. And quite a little story goes with that. I hesitate to tell it. It is like opening an old sore. My tongue seems to swell and my throat to crack when I think of it now. Other people have gone many days without food or water, but I never realized how they suffered until I was forced to go two days without those things myself.

When we finally did get started that morning from the trail below the cabin, we moved off splendidly. The cowboys were there to see the start; Davis and Pitts and the Mexican, all mounted on fiery broncos that snorted and wheeled every time the cylinders puffed out a little steam. But the cowboys sat in the saddles like they were part of the broncos.

For ten miles or more we scooted through the forest like sliding down the chutes. It was a glorious morning and a glorious ride. Twice we came close to bands of antelope, beautiful, graceful creatures, whose curiosity to know what made the automobile go kept them from running away until we were nearly upon them. Once we saw a number of wild horses way up on the side of a hill. They were looking for snow to eat, for they were many miles from water.

We had just descended a small hill where the trail was rough with stones, and the driver had given her more steam as we struck the level and good going again, when something popped, and instantly we were enveloped in a cloud of steam. Our water gauge had burst, and the valves had failed to work. We lost all our steam, and, I am afraid, our hope of heaven, right there. And that was only the beginning of our troubles that day.

We had now used all of the gasoline brought from Los Angeles, which had been contained in the two feed tanks under the seat, and were obliged to refill the tanks with oil purchased in Flagstaff, and alleged to be gasoline. When we started again it was plain to be seen that this fuel was not going to give the amount of heat required to keep up sufficient steam. Besides, it produced a dense volume of smoke, that poured out of the ventilators and enveloped us, making us look like Indians. We crawled along for several hours, making about three miles an hour. Then we decided to abandon the trailer, with its load of baggage, water and alleged gasoline, and attempt to make the canyon some time during the night. So everything not absolutely necessary was left right there, and all four of us piled onto the machine in a heap, and she started. She started well, and kept up a good rate of speed for a mile or more. Then, all at once, we heard a sharp, metallic click, and in another instant a harsh, rasping sound, and we knew that our sprocket chain had parted. An automobile always gets discouraged and quits when its sprocket chain parts.

## A Night Under the Stars

Three hours later the chain was mended. It was now nearly 12 o'clock at night. We were out on the open plain. The night was bitter cold, and the wind blew right up under our coat tails. Doyle reckoned there was a bunch of cedars about four miles ahead where we could find shelter. We could see a dark fringe way off in the dim horizon. The kerosene oil in our side lamps was exhausted and they shed a pale, sickly light only a few feet ahead. But after two hours more, with two of us walking ahead to discover the way, we pulled into the bunch of cedars which Doyle had reckoned was there. Here we built two huge campfires, and after a hearty supper of vain regrets and some delicious thoughts of home, we laid down to sleep. At least that was what we intended to do. What we did was to shiver for half an hour or so, and then get up and pile more wood on the fires, and smoke.

At daylight we were up and ready for breakfast. For breakfast we each had a look at the automobile and a smoke. Not one of us said anything about being hungry or even thirsty. There was a little dirty ice in the bottom of the tank and we melted some of it, but it didn't taste good so soon after breakfast. So we took another smoke, and let it go at that.

The inspection of the automobile revealed the fact that there was just a little gasoline left in the tanks and a little water in the boiler. Doyle reckoned it was about eighteen miles to Pete Berry's.

## The Auto Abandoned

We started the fire under the boilers and finally got up a little steam. Chauffeur Lippincott worried the auto along for about two miles and the rest of us walked. Finally, at the end of the two miles, we were obliged to abandon the machine.

We were again in a great pine forest. We had no trail to follow, but depended on Doyle. He reckoned he could get us there. We had discarded every bit of clothing that weighed anything worth considering, and were plodding along, and saying very little. Eighteen miles isn't such a long walk. But we were hungry and thirsty and weak from lack of sleep and other things.

At noon we sighted Skinner's cabin. It is an old log cabin, with a legend attached to it to the effect that a man by the name of Skinner built it, years ago. Skinner must have been crazy. There isn't any more call for a cabin there than there is for a department store. On the cabin was a sign which set forth that it was 6-3/8 miles to the Grand canyon. The man who put up that sign ought to be killed with a dull hatchet.

Two members of the party, Chauffeur Lippincott and Journalist Chapman, were petered out. They could go no further. Doyle and myself agreed to make the remaining distance to Berry's and send out a relief expedition. Earlier in the day Lippincott had stated positively that he would give $5 for a drink of water, and by degrees he had raised the amount until it now stood at $500. Chapman had offered $50 for a ham sandwich, with no takers, and Doyle reckoned he'd be willing to give more than that for a good steak if he had it.

## On the Home Stretch

Doyle's stomach wasn't acting right. It probably thought his throat had been cut, or something of the sort. The two of us had covered scarcely half a mile when he collapsed utterly, and gave up the struggle.

Now it was after 1 o-clock, and I was hungry; also thirsty; also tired—O, so tired! I lay down on a little pile of nice, soft lumps of lava to rest for just a minute, and fell asleep. I dreamed of home; of the dinner table. There was a big fat turkey before me, with dressing and cranberry sauce. It was terribly realistic. I was just going to feed myself a very generous piece of the white meat when Doyle woke me.

I left him there by the lava rocks, wrestling with his stomach, and went on alone.

The way was still through the great forest. The silence was oppressive. I could hear the murmur of distant voices now and then, and the rippling of a brook far away. Of course, there were not any voices or any brook. There wasn't anything but trees, and the blue sky.

Once I heard a bear coming behind me. I hated to be bothered with bears at that time, but I resolved to sell my life as dearly as possible, and decided to wait until he was real close to me and then take my jack-knife in my hand, thrust it quickly into the bear's mouth and cut his throat on the inside. Luckily for me, however, it turned out not to be a bear at all; simply a pine cone dropping from a tree.

After I had gone about eight miles I happened to think of that sign on Skinner's cabin, "Six and three-eighths miles to the Grand Canyon." What the deuce did the man want to be so exact for? If he was going to lie about it, why didn't he lie in round numbers, at least? If he had said, "Six miles to the Grand Canyon," or "About six miles to the Grand Canyon," I might have felt differently towards him; but for him to be confoundedly exact about it, when he knew that it wasn't an inch less than twelve miles, with the chances better that it was about fourteen, made me exasperated. I resolved right there to find the man who put up that sign if it took half a lifetime, which is all and perhaps more than I've got left, and to kill him without mercy.

I didn't dare to stop. If I stopped I would fall asleep and get to dreaming about turkey and cranberry sauce again. "Six and three-eighths miles to the Grand Canyon!" If that man had appeared there before me I would have stopped long enough to have killed him, at least. My tongue had swollen until it made a good big mouthful, and my throat was cracking. I tried to talk to myself, so as not to lose my voice. I wanted to be able to inquire the whereabouts of that man when I got to Berry's.

About 4 o'clock, after I had covered about eighteen miles of that "six and three-eighths miles to the Grand Canyon," I noticed something moving at a spot perhaps a quarter of a mile away, in the bottom of a little canyon. At first I thought it might be the man who put up that sign, and my heart leaped for joy. But after a long look I made out the object to be a big buck. He was pawing something white on the ground. It was a small patch of snow! I climbed up the little canyon, sat down beside the snow and ate up about two quarts of it. Then I gathered up a lot more and filled all my pockets with it. That snow tasted better than the most delicious ice cream I ever ate. It was not very clean, for the deer had trampled it all over but I didn't mind that.

An hour later, as I looked ahead, I could see a break in the trees. I was on the edge of a clearing of some kind, anyhow, and the terrible monotony and awful silence of the forest were to be broken at least. With renewed energy I wobbled on and came to a little hill, beyond the summit of which there were no trees. I had grown to hate trees.

67

Slowly I crawled up the little hill and stood upon its crest. Then I saw why there were no trees beyond the hill. Beyond the hill was the grandest, most awe-inspiring sight in all the world—the Grand Canyon of the Colorado.

I stood there upon the rim of that tremendous chasm and forgot who I was, and what I came there for. Before me lay the sublimest panorama in the world. Nature never made anything like it anywhere else. It is the great masterpiece, and as I viewed it, and my eyes drank in the grandeur of the scene, I said to myself, "You better stand back a little further; you might fall in."

Twenty minutes later I was at the Grand View hotel and Landlord Berry was asking me what I wanted.

"All I want is water and food, and the address of the man who put up the sign on Skinner's cabin," I said, modestly.

They brought me a little water and a can of tomatoes, but the address they withheld. Possibly the man was a friend, or he may be the sole support of a large family.

That night, about 8 o'clock, the relief expedition, consisting of Landlord Berry and a four-horse team, returned to the hotel on the rim, bringing Chauffeur Lippincott, Journalist Chapman and Guide Doyle, the sole survivors. At 9 o'clock we ate enough to have driven any other landlord except Pete Berry into fits, and at 10 o'clock we retired.

Next day Berry and I drove over to Bright Angel hotel and telephoned for gasoline. The day after that the gasoline came, and the day after that day, which was Thursday, we got the automobile to the Grand canyon.

Friday, we steamed up and took the auto out on Grand View point. Chauffeur Lippincott drove the thing to within six inches of the rim with its own steam and held it there while I took its picture.

Saturday morning before daylight Doyle, Chapman and myself drove over to Apex, sixteen miles from Grand View, and took the train for home. Every one of us regretted that we would be unable to make the return trip from the Grand Canyon to Flagstaff, but the facts in the case were that Doyle was obliged to get back to Flagstaff in order to assist his wife in setting a hen, and Chapman had to be in Los Angeles in time to prepare a speech for the banquet of the Typothetae on Franklin's birthday, while in my own case it was absolutely imperative that I be at home to pay some bills that had accumulated during my absence.

Several days later Mr. Lippincott made the return trip from the Grand View hotel to Flagstaff, sixty-seven miles, in seven hours.

Without doubt, an automobile line will be established next summer between Flagstaff and the canyon. It is the best, and in fact the only route for sightseers. The country is full of natural attractions, and the scenery and climate are superb.

I am still searching for the man who put up that sign – "Six and three-eighths miles to the Grand canyon" – but the chances are that if some one comes along and asks me to join a party to make an experimental trip in a flying machine from Grubb's Corners to Patagonia I'll go; I'm just that foolish.

CHAPTER FIVE

# 1902
# Our Grand Canyon

## John Muir

John Muir, one of America's first important proponents of
wilderness preservation, has achieved a stature in conservation
history that approaches the prophetic. His greatest influence
was through his writings, and his name has always been most
closely associated with Yosemite National Park in California.
In this article he even rises to defend Yosemite against those
who would make the apples-and-oranges comparison of the
Yosemite Valley with the Grand Canyon.

Muir's writing could just as easily be preached as read. Giant
canyons and towering peaks were his home territory, and his
prose is filled with monumental imagery and colors of almost
roaring brilliance. The powers of nature at their grandest—
thundering storms, irresistible glaciers, and excavating
rivers—were his favorite dramatic characters, so the Grand
Canyon was just his kind of literary subject.

H APPY nowadays is the tourist, with earth's wonders, new
and old, spread invitingly open before him, and a host of
able workers as his slaves, making everything easy, padding plush

about him, grading roads for him, boring tunnels, moving hills out of his way, eager, like the devil, to show him all the kingdoms of the world and their glory and foolishness, spiritualizing travel for him with lightning and steam, abolishing space and time and almost everything else. Little children and tender, pulpy people, as well as storm-seasoned explorers, may now go almost everywhere in smooth comfort, cross oceans and deserts scarce accessible to fishes and birds, and, dragged by steel horses, go up high mountains, riding gloriously beneath starry showers of sparks, ascending like Elijah in a whirlwind and chariot of fire.

First of the wonders of the great West to be brought within reach of the tourist were the Yosemite and the Big Trees, on the completion of the first transcontinental railway; next came the Yellowstone and icy Alaska, by the Northern roads; and last the Grand Cañon of the Colorado, which, naturally the hardest to reach, has now become, by a branch of the Santa Fé, the most accessible of all.

Of course with this wonderful extension of steel ways through our wildness there is loss as well as gain. Nearly all railroads are bordered by belts of desolation. The finest wilderness perishes as if stricken with pestilence. Bird and beast people, if not the dryads, are frightened from the groves. Too often the groves also vanish, leaving nothing but ashes. Fortunately, nature has a few big places beyond man's power to spoil—the ocean, the two icy ends of the globe, and the Grand Cañon.

When I first heard of the Santa Fé trains running to the edge of the Grand Cañon of the Arizona, I was troubled with thoughts of the disenchantment likely to follow. But last winter, when I saw those trains crawling along through the pines of the Cocanini Forest and close up to the brink of the chasm at Bright Angel, I was glad to discover that in the presence of such stupendous scenery they are nothing. The locomotives and trains are mere beetles and caterpillars, and the noise they make is as little disturbing as the hooting of an owl in the lonely woods.

In a dry, hot, monotonous forested plateau, seemingly boundless, you come suddenly and without warning upon the abrupt edge of a gigantic sunken landscape of the wildest, most multitudinous features, and those features, sharp and angular, are made out of flat beds of limestone and sandstone forming a spiry, jagged, gloriously colored mountain-range countersunk in a level gray plain. It is a hard job to sketch it even in scrawniest outline; and try as I may, not in the least sparing myself, I cannot tell the hundredth part of the wonders of its features—the side-cañons, gorges, alcoves, cloisters, and amphitheaters of vast sweep and depth, carved in its magnificent walls; the throng of great architectural rocks it contains resembling castles, cathedrals, temples, and palaces, towered and spired and painted, some of them nearly a mile high, yet beneath one's feet. All this, however, is less difficult than to give any idea of the impression of wild, primeval beauty and power one receives in merely gazing from its brink. The view down the gulf of color and over the rim of its wonderful wall, more than any other view I know, leads us to think of our earth as a star with stars swimming in light, every radiant spire pointing the way to the heavens.

But it is impossible to conceive what the cañon is, or what impression it makes, from descriptions or pictures, however good. Naturally it is untellable even to those who have seen something perhaps a little like it on a small scale in this same plateau region. One's most extravagant expectations are indefinitely surpassed, though one expects much from what is said of it as "the biggest chasm on earth"—"so big is it that all other big things—Yosemite, the Yellowstone, the Pyramids, Chicago—all would be lost if tumbled in it." Naturally enough, illustrations as to size are sought for among other cañons like or unlike it, with the common result of worst confounding confusion. The prudent keep silence. It was once said that the "Grand Cañon could put a dozen Yosemites in its vest pocket."

The justly famous Grand Cañon of the Yellowstone is, like the Colorado, gorgeously colored and abruptly countersunk in a plateau, and both are mainly the work of water. But the Colorado's cañon is more than a thousand times larger, and as a score or two new buildings of ordinary size would not appreciably change the general view of a great city, so hundreds of Yellowstones might be eroded in the sides of the Colorado Cañon without noticeably augmenting its size or the richness of its sculpture. But it is not true that the great Yosemite rocks would be thus lost or hidden. Nothing of their kind in the world, so far as I know, rivals El Capitan and Tissiack, much less dwarfs or in any way belittles them. None of the sandstone or limestone precipices of the cañon that I have seen or heard of approaches in smooth, flawless strength and grandeur the granite face of El Capitan or the Tenaya side of Cloud's Rest. These colossal cliffs, types of permanence, are about three thousand and six thousand feet high; those of the cañon that are sheer are about half as high, and are types of fleeting change; while glorious-domed Tissiack, noblest of mountain buildings, far from being overshadowed or lost in this rosy, spiry cañon company, would draw every eye, and, in serene majesty, "aboon them a'" she would take her place—castle, temple, palace, or tower. Nevertheless a noted writer, comparing the Grand Cañon in a general way with the glacial Yosemite, says: "And the Yosemite—ah, the lovely Yosemite! Dumped down into the wilderness of gorges and mountains, it would take a guide who knew of its existence a long time to find it." This is striking, and shows up well above the levels of commonplace description; but it is confusing, and has the fatal fault of not being true. As well try to describe an eagle by putting a lark in it. "And the lark—ah, the lovely lark! Dumped down the red, royal gorge of the eagle, it would be hard to find." Each in its own place is better, singing at heaven's gate, and sailing the sky with the clouds.

Every feature of nature's big face is beautiful—height and hollow, wrinkle, furrow, and line—and this is the main master furrow of its kind on our continent, incomparably greater and more impressive than any other yet discovered, or likely to be dis-covered, now that all the great rivers have been traced to their heads.

The Colorado River rises in the heart of the continent on the dividing ranges and ridges between the two oceans, drains thou-sands of snowy mountains through narrow or spacious valleys, and thence through cañons of every color, sheer-walled and deep, all of which seem to be represented in this one grand cañon of cañons.

It is very hard to give anything like an adequate conception of its size, much more of its color, its vast wall-sculpture, the wealth of ornate architectural buildings that fill it, or, most of all, the tre-mendous impression it makes. According to Major Powell, it is about two hundred and seventeen miles long, from five to fifteen miles wide from rim to rim, and from about five thousand to six thousand feet deep. So tremendous a chasm would be one of the world's greatest wonders even if, like ordinary cañons cut in sedimentary rocks, it were empty and its walls were simple. But instead of being plain, the walls are so deeply and elaborately carved into all sorts of recesses—alcoves, cirques, amphitheaters, and side cañons—that were you to trace the rim closely around on both sides your journey would be nearly a thousand miles long. Into all these recesses the level, continuous beds of rock in ledges and benches, with their various colors, run like broad ribbons, marvelously beautiful and effective even at a distance of ten or twelve miles. And the vast space these glorious walls inclose, instead of being empty, is crowded with gigantic architectural rock forms gorgeously colored and adorned with towers and spires like works of art.

Looking down from this level plateau, we are more impressed with a feeling of being on the top of everything than when looking from the summit of a mountain. From side to side of the vast gulf, temples, palaces, towers, and spires come soaring up in thick array half a mile or nearly a mile above their sunken, hidden bases, some to a level with our standpoint, but none higher. And in the inspiring morning light all are so fresh and rosy-looking that they seem new-born; as if, like the quick-growing crimson snow-plants of the California woods, they had just sprung up, hatched by the warm, brooding, motherly weather.

In trying to describe the great pines and sequoias of the Sierra, I have often thought that if one of those trees could be set by itself in some city park, its grandeur might there be impressively realized; while in its home forests, where all magnitudes are great, the weary, satiated traveler sees none of them truly. It is so with these majestic rock structures.

Though mere residual masses of the plateau, they are dowered with the grandeur and repose of mountains, together with the finely chiseled carving and modeling of man's temples and palaces, and often, to a considerable extent, with their symmetry. Some, closely observed, look like ruins; but even these stand plumb and true, and show architectural forms loaded with lines strictly regular and decorative, and all are arrayed in colors that storms and time seem only to brighten. They are not placed in regular rows in line with the river, but "a' through ither," as the Scotch say, in

75

Theodore Roosevelt and John Hance, first and second in line, coming down Bright Angel Trail. *Photo from the Emery Kolb Collection, Northern Arizona University Libraries. Copyright 1981 by Emery Lehnert*

lavish, exuberant crowds, as if nature in wildest extravagance held her bravest structures as common as gravel-piles. Yonder stands a spiry cathedral nearly five thousand feet in height, nobly symmetrical, with sheer buttressed walls and arched doors and windows, as richly finished and decorated with sculptures as the great rock temples of India or Egypt. Beside it rises a huge castle with arched gateway, turrets, watch-towers, ramparts, etc., and to right and left palaces, obelisks, and pyramids fairly fill the gulf, all colossal and all lavishly painted and carved. Here and there a flat-topped structure may be seen, or one imperfectly domed; but the prevailing style is ornate Gothic, with many hints of Egyptian and Indian.

Throughout this vast extent of wild architecture—nature's own capital city—there seem to be no ordinary dwellings. All look like grand and important public structures, except perhaps some of the lower pyramids, broad-based and sharp-pointed, covered with down-flowing talus like loosely set tents with hollow, sagging sides. The roofs often have disintegrated rocks heaped and draggled over them, but in the main the masonry is firm and laid in regular courses, as if done by square and rule.

Nevertheless they are ever changing: their tops are now a dome, now a flat table or a spire, as harder or softer strata are reached in their slow degradation, while the sides, with all their fine moldings, are being steadily undermined and eaten away. But no essential change in style or color is thus effected. From century to century they stand the same. What seems confusion among the rough earthquake-shaken crags nearest one comes to order as soon as the main plan of the various structures appears. Every building, however complicated and laden with ornamental lines, is at one with itself and every one of its neighbors, for the same characteristic controlling belts of color and solid strata extend with wonderful constancy for very great distances, and pass through and give style to thousands of separate structures, however their smaller characters may vary.

Of all the various kinds of ornamental work displayed—carving, tracery on cliff-faces, moldings, arches, pinnacles—none is more admirably effective or charms more than the webs of rain-channeled taluses. Marvelously extensive, without the slightest appearance of waste or excess, they cover roofs and dome-tops and the base of every cliff, belt each spire and pyramid and massy, tower-ing temple, and in beautiful continuous lines go sweeping along the great walls in and out around all the intricate system of side-cañons, amphitheaters, cirques, and scallops into which they are sculptured. From one point hundreds of miles of this fairy em-broidery may be traced. It is all so fine and orderly that it would seem that not only had the clouds and streams been kept har-moniously busy in the making of it, but that every raindrop sent like a bullet to a mark had been the subject of a separate thought, so sure is the outcome of beauty through the stormy centuries. Surely nowhere else are there illustrations so striking of the nat-ural beauty of desolation and death, so many of nature's own mountain buildings wasting in glory of high desert air—going to dust. See how steadfast in beauty they all are in their going. Look again and again how the rough, dusty boulders and sand of disinte-gration from the upper ledges wreathe in beauty the next and next below with these wonderful taluses, and how the colors are finer the faster the waste. We oftentimes see nature giving beauty for ashes—as in the flowers of a prairie after fire—but here the very dust and ashes are beautiful.

Gazing across the mighty chasm, we at last discover that it is not its great depth nor length, nor yet these wonderful buildings, that most impresses us. It is its immense width, sharply defined by pre-cipitous walls plunging suddenly down from a flat plain, declaring in terms instantly apprehended that the vast gulf is a gash in the once unbroken plateau, made by slow, orderly erosion and removal of huge beds of rocks. Other valleys of erosion are as great—in all their dimensions some are greater—but none of these produces an effect on the imagination at once so quick and pro-found, coming without study, given at a glance. Therefore by far the greatest and most influential feature of this view from Bright Angel or any other of the cañon views is the opposite wall. Of the one beneath our feet we see only fragmentary sections in cirques and amphitheaters and on the sides of the outjutting promontories between them, while the other, though far distant, is beheld in all its glory of color and noble proportions—the one supreme beauty and wonder to which the eye is ever turning. For while charming with its beauty it tells the story of the stupendous erosion of the cañon—the foundation of the unspeakable impression made on

78

everybody. It seems a gigantic statement for even nature to make, all in one mighty stone word, apprehended at once like a burst of light, celestial color its natural vesture, coming in glory to mind and heart as to a home prepared for it from the very beginning. Wildness so godful, cosmic, primeval, bestows a new sense of earth's beauty and size. Not even from high mountains does the world seem so wide, so like a star in glory of light on its way through the heavens.

I have observed scenery-hunters of all sorts getting first views of Yosemites, glaciers, White Mountain ranges, etc. Mixed with the enthusiasm which such scenery naturally excites, there is often weak gushing, and many splutter aloud like little waterfalls. Here, for a few moments at least, there is silence, and all are in dead earnest, as if awed and hushed by an earthquake—perhaps until the cook cries "Breakfast!" or the stable-boy "Horses are ready!" Then the poor unfortunates, slaves of regular habits, turn quickly away, gasping and muttering as if wondering where they had been and what had enchanted them.

Roads have been made from Bright Angel Hotel through the Cocanini Forest to the ends of outstanding promontories, commanding extensive views up and down the cañon. The nearest of them, three or four miles east and west, are McNeil's Point and Rowe's Point; the latter, besides commanding the eternally interesting cañon, gives wide-sweeping views southeast and west over the dark forest roof to the San Francisco and Mount Trumbull volcanoes—the bluest of mountains over the blackest of level woods.

Instead of thus riding in dust with the crowd, more will be gained by going quietly afoot along the rim at different times of day and night, free to observe the vegetation, the fossils in the rocks, the seams beneath overhanging ledges once inhabited by Indians, and to watch the stupendous scenery in the changing lights and shadows, clouds, showers, and storms. One need not go hunting the so-called "points of interest." The verge anywhere, everywhere, is a point of interest beyond one's wildest dreams.

As yet, few of the promontories or throng of mountain buildings in the cañon are named. Nor among such exuberance of forms are names thought of by the bewildered, hurried tourist. He would be as likely to think of names for waves in a storm. The

Eastern and Western Cloisters, Hindu Amphitheater, Cape Royal, Powell's Plateau, and Grand View Point, Point Sublime, Bissell and Moran points, the Temple of Set, Vishnu's Temple, Shiva's Temple, Twin Temples, Tower of Babel, Hance's Column—these fairly good names given by Dutton, Holmes, Moran, and others are scattered over a large stretch of the cañon wilderness.

All the cañon rock-beds are lavishly painted, except a few neutral bars and the granite notch at the bottom occupied by the river, which makes but little sign. It is a vast wilderness of rocks in a sea of light, colored and glowing like oak and maple woods in autumn, when the sungold is richest. I have just said that it is impossible to learn what the cañon is like from descriptions and pictures. Powell's and Dutton's descriptions present magnificent views not only of the cañon but of all the grand region round about it; and Holmes's drawings, accompanying Dutton's report, are wonderfully good. Surely faithful and loving skill can go no further in putting the multitudinous decorated forms on paper. But the *colors,* the living, rejoicing *colors,* chanting morning and evening in chorus to heaven! Whose brush or pencil, however lovingly inspired, can give us these? And if paint is of no effect, what hope lies in pen-work? Only this: some may be incited by it to go and see for themselves.

No other range of mountainous rockwork of anything like the same extent have I seen that is so strangely, boldly, lavishly colored. The famous Yellowstone Cañon below the falls comes to mind; but, wonderful as it is, and well deserved as is its fame, compared with this it is only a bright rainbow ribbon at the roots of the pines. Each of the series of level, continuous beds of carboniferous rocks of the cañon has, as we have seen, its own characteristic color. The summit limestone-beds are pale yellow; next below these are the beautiful rose-colored cross-bedded sandstones; next there are a thousand feet of brilliant red sandstones; and below these the red wall limestones, over two thousand feet thick, rich massy red, the greatest and most influential of the series, and forming the main color-fountain. Between these are many neutral-tinted beds. The prevailing colors are wonderfully deep and clear, changing and blending with varying intensity from hour to hour, day to day, season to season; throbbing, wavering, glowing, responding to every passing cloud or storm, a world of color in itself, now burning in separate rainbow bars streaked and blotched with shade, now glowing in one smooth, all-pervading ethereal radiance like the alpenglow, uniting the rocky world with the heavens.

The dawn, as in all the pure, dry desert country, is ineffably beautiful; and when the first level sunbeams sting the domes and spires, with what a burst of power the big, wild days begin! The dead and the living, rocks and hearts alike, awake and sing the new-old song of creation. All the massy headlands and salient angles of the walls, and the multitudinous temples and palaces, seem to catch the light at once, and cast thick black shadows athwart hollow and gorge, bringing out details as well as the main massive features of the architecture; while all the rocks, as if wild with life, throb and quiver and glow in the glorious sunburst, rejoicing. Every rock temple then becomes a temple of music; every spire and pinnacle an angel of light and song, shouting color halleluiahs.

As the day draws to a close, shadows, wondrous, black, and thick, like those of the morning, fill up the wall hollows, while the glowing rocks, their rough angles burned off, seem soft and hot to the heart as they stand submerged in purple haze, which now fills the cañon like a sea. Still deeper, richer, more divine grow the great walls and temples, until in the supreme flaming glory of sunset the whole cañon is transfigured, as if all the life and light of centuries of sunshine stored up and condensed in the rocks was now being poured forth as from one glorious fountain, flooding both earth and sky.

Strange to say, in the full white effulgence of the midday hours the bright colors grow dim and terrestrial in common gray haze; and the rocks, after the manner of mountains, seem to crouch and drowse and shrink to less than half their real stature, and have nothing to say to one, as if not at home. But it is fine to see how quickly they come to life and grow radiant and communicative as soon as a band of white clouds come floating by. As if shouting for joy, they seem to spring up to meet them in hearty salutation, eager to touch them and beg their blessings. It is just in the midst of these dull midday hours that the cañon clouds are born.

A good storm-cloud full of lightning and rain on its way to its work on a sunny desert day is a glorious object. Across the cañon, opposite the hotel, is a little tributary of the Colorado called Bright Angel Creek. A fountain-cloud still better deserves the name "Angel of the Desert Wells"—clad in bright plumage, carrying cool shade and living water to countless animals and plants ready to perish, noble in form and gesture, seeming able for anything, pouring life-giving, wonder-working floods from its alabaster fountains, as if some sky-lake had broken. To every gulch and gorge on its favorite ground is given a passionate torrent, roaring, replying to the rejoicing lightning—stones, tons in weight, hurrying away as if frightened, showing something of the way Grand Cañon work is done. Most of the fertile summer clouds of the cañon are of this sort, massive, swelling cumuli, growing rapidly, displaying delicious tones of purple and gray in the hollows of their sun-beaten bosses, showering favored areas of the heated landscape, and vanishing in an hour or two. Some, busy and thoughtful-looking, glide with beautiful motion along the middle of the cañon in flocks, turning aside here and there, lingering as if studying the needs of particular spots, exploring side-cañons, peering into hollows like birds seeking nest-places, or hovering aloft on outspread wings. They scan all the red wilderness, dispensing their blessings of cool shadows and rain where the need is the greatest, refreshing the rocks, their offspring as well as the vegetation, continuing their sculpture, deepening gorges and sharpening peaks. Sometimes, blending all together, they weave a ceiling from rim to rim, perhaps opening a window here and there for sunshine to stream through, suddenly lighting some palace or temple and making it flare in the rain as if on fire.

Sometimes, as one sits gazing from a high, jutting promontory, the sky all clear, showing not the slightest wisp or penciling, a bright band of cumuli will appear suddenly, coming up the cañon in single file, as if tracing a well-known trail, passing in review, each in turn darting its lances and dropping its shower, making a row of little vertical rivers in the air above the big brown one. Others seem to grow from mere points, and fly high above the cañon, yet following its course for a long time, noiseless, as if hunting, then suddenly darting lightning at unseen marks, and hurrying on. Or they loiter here and there as if idle, like laborers out of work, waiting to be hired.

Half a dozen or more showers may oftentimes be seen falling at once, while far the greater part of the sky is in sunshine, and not a raindrop comes nigh one. These thundershowers from as many separate clouds, looking like wisps of long hair, may vary greatly in effects. The pale, faint streaks are showers that fail to reach the ground, being evaporated on the way down through the dry, thirsty air, like streams in deserts. Many, on the other hand, which in the distance seem insignificant, are really heavy rain, however local; these are the gray wisps well zigzagged with lightning. The darker ones are torrent rain, which on broad, steep slopes of favorable conformation give rise to so-called "cloudbursts"; and wonderful is the commotion they cause. The gorges and gulches below them, usually dry, break out in loud uproar, with a sudden downrush of muddy, boulder-laden floods. Down they all go in one simultaneous gush, roaring like lions rudely awakened, each of the tawny brood actually kicking up a dust at the first onset.

During the winter months snow falls over all the high plateau, usually to a considerable depth, whitening the rim and the roofs of the cañon buildings. But last winter, when I arrived at Bright Angel in the middle of January, there was no snow in sight, and the ground was dry, greatly to my disappointment, for I had made the trip mainly to see the cañon in its winter garb. Soothingly I was informed that this was an exceptional season, and that the good snow might arrive at any time. After waiting a few days, I gladly hailed a broad-browed cloud coming grandly on from the west in big promising blackness, very unlike the white sailors of the summer skies. Under the lee of a rim-ledge, with another snow-lover, I watched its movements as it took possession of the cañon and all the adjacent region in sight. Trailing its gray fringes over the spiry tops of the great temples and towers, it gradually settled lower, embracing them all with ineffable kindness and gentleness of touch, and fondled the little cedars and pines as they quivered eagerly in the wind like young birds begging their mothers to feed them. The first flakes and crystals began to fly about noon, sweeping straight up the middle of the cañon, and swirling in magnificent eddies along the sides. Gradually the hearty swarms closed their ranks, and all the cañon was lost in gray gloom except a short section of the wall and a few trees beside us, which looked glad with snow in their needles and about their feet as they leaned out over the gulf. Suddenly the storm opened with magical effect to the north over the cañon of Bright Angel Creek, inclosing a sunlit mass of the cañon architecture, spanned by great white concentric arches of cloud like the bows of a silvery aurora. Above these and a little back of them was a series of upboiling purple clouds, and high above all, in the background, a range of noble cumuli towered aloft like snow-laden mountains, their pure pearl bosses flooded with sunshine. The whole noble picture, calmly glowing, was framed in thick gray gloom, which soon closed over it; and the storm went on, opening and closing until night covered all.

Two days later, when we were on a jutting point about eighteen miles east of Bright Angel and one thousand feet higher, we enjoyed another storm of equal glory as to cloud effects, though only a few inches of snow fell. Before the storm began we had a magnificent view of this grander upper part of the cañon and also of the Cocanini Forest and Painted Desert. The march of the clouds with their storm-banners flying over this sublime land-scape was unspeakably glorious, and so also was the breaking up of the storm next morning—the mingling of silver-capped rock, sunshine, and cloud.

Most tourists make out to be in a hurry even here; therefore their few days or hours would be best spent on the promontories nearest the hotel. Yet a surprising number go down the Bright Angel trail to the brink of the inner gloomy granite gorge over-looking the river. Deep cañons attract like high mountains; the deeper they are, the more surely are we drawn into them. On foot, of course, there is no danger whatever, and, with ordinary precautions, but little on animals. In comfortable tourist faith, unthinking, unfearing, down go men, women, and children on whatever is offered, horse, mule, or burro, as if saying with Jean Paul, "fear nothing but fear"—not without reason, for these cañon trails down the stairways of the gods are less dangerous than they seem, less dangerous than home stairs. The guides are cautious, and so are the experienced, much-enduring beasts: The scrawniest Rosinantes and wizened-rat mules cling hard to the rocks endwise or sidewise, like lizards or ants. From terrace to terrace, climate to climate, down one creeps in sun and shade, through gorge and gully and grassy ravine, and, after a long scramble on foot, at last beneath the mighty cliffs one comes to the grand, roaring river.

To the mountaineer the depth of the cañon, from five thousand to six thousand feet, will not seem so very wonderful, for he has often explored others that are about as deep. But the most experienced will be awe-struck by the vast extent of strange, countersunk scenery, the multitude of huge rock monuments of painted masonry built up in regular courses towering above, beneath, and round about him. By the Bright Angel trail the last fifteen hundred feet of the descent to the river has to be made afoot down the gorge of Indian Garden Creek. Most of the visitors do not like this part, and are content to stop at the end of the horse-trail and look down on the dull-brown flood from the edge of the Indian Garden Plateau. By the new Hance trail, excepting a few daringly steep spots, you can ride all the way to the river, where there is a good spacious camp-ground in a mesquite-grove. This trail, built by brave Hance, begins on the highest part of the rim, eight thousand feet above the sea, a thousand feet higher than the head of Bright Angel trail, and the descent is a little over six thousand feet, through a wonderful variety of climate and life. Often late in the fall, when frosty winds are blowing and snow is flying at one end of the trail, tender plants are blooming in balmy summer weather at the other. The trip down and up can be made afoot easily in a day. In this way one is free to observe the scenery and vegetation, instead of merely clinging to his animal and watching its steps. But all who have time should go prepared to camp awhile on the riverbank, to rest and learn something about the plants and animals and the mighty flood roaring past. In cool, shady amphitheaters at the head of the trail there are groves of white silver fir and Douglas spruce, with ferns and saxifrages that recall snowy mountains; below these, yellow pine, nut-pine, juniper, hop-hornbeam, ash, maple, holly-leaved berberis, cowania, spiræa, dwarf oak, and other small shrubs and trees. In dry gulches and on taluses and sun-beaten crags are sparsely scattered yuccas, cactuses, agave, etc. Where springs gush from the rocks there are willow thickets, grassy flats, and bright flowery gardens, and in the hottest recesses the delicate abronia, mesquite, woody compositæ, and arborescent cactuses.

The most striking and characteristic part of this widely varied vegetation are the cactaceæ—strange, leafless, old-fashioned plants with beautiful flowers and fruit, in every way able and admirable. While grimly defending themselves with innumerable barbed spears, they offer both food and drink to man and beast. Their juicy gloves and disks and fluted cylindrical columns are almost the only desert wells that never go dry, and they always seem to rejoice the more and grow plumper and juicier the hotter the sunshine and sand. Some are spherical, like rolled-up porcupines, crouching in rock hollows beneath a mist of gray lances, unmoved by the wildest winds. Others, standing as erect as bushes and trees or tall branchless pillars crowned with magnificent flowers, their prickly armor sparkling, look boldly abroad over the glaring desert, making the strangest forests ever seen or dreamed of. *Cereus giganteus,* the grim chief of the desert tribe, is often thirty or forty feet high in southern Arizona. Several species of tree yuccas in the same deserts, laden in early spring with superb white lilies, form forests hardly less wonderful, though here they grow singly or in small lonely groves. The low, almost stemless *Yucca baccata,* with beautiful lily-flowers and sweet banana-like fruit, prized by the Indians, is common along the cañon rim, growing on lean, rocky soil beneath mountain-mahogany, nut-pines, and junipers, beside dense flowery mats of *Spiræa cæspitosa* and the beautiful pinnate-leaved *Spiræa millefolium.* The nut-pine, *Pinus edulis,* scattered along the upper slopes and roofs of the cañon buildings, is the principal tree of the strange Dwarf Cocanini Forest. It is a picturesque stub of a pine about twenty-five feet high, usually with dead, lichened limbs thrust through its rounded head, and grows on crags and fissured rock tables, braving heat and frost, snow and drought, and continues patiently, faithfully fruitful for centuries. Indians and insects and almost every desert bird and beast come to it to be fed.

To civilized people from corn and cattle and wheat-field countries the cañon at first sight seems as uninhabitable as a glacier crevasse, utterly silent and barren. Nevertheless it is the home of a multitude of our fellow-mortals, men as well as animals and plants. Centuries ago it was inhabited by tribes of Indians, who, long before Columbus saw America, built thousands of stone houses in its crags, and large ones, some of them several stories high, with hundreds of rooms, on the mesas of the adjacent regions. Their cliff-dwellings, almost numberless, are still to be seen in the cañon, scattered along both sides from top to bottom and throughout its entire length, built of stone and mortar in seams and fissures like swallows' nests, or on isolated ridges and peaks. The ruins of larger buildings are found on open spots by the river, but most of them aloft on the brink of the wildest, giddiest precipices, sites evidently chosen for safety from enemies, and seemingly accessible only to the birds of the air. Many caves were also used as dwelling-places, as were mere seams on cliff-fronts formed by unequal weathering and with or without outer or side walls; and some of them were covered with colored pictures of animals. The most interesting of these cliff-dwellings had pathetic little ribbon-like strips of garden on narrow terraces, where irrigating-water could be carried to them—most romantic of sky-gardens, but eloquent of hard times.

In recesses along the river and on the first plateau flats above its gorge were fields and gardens of considerable size, where irrigating ditches may still be traced. Some of these ancient gardens are still cultivated by Indians, descendants of cliff dwellers, who raise corn, squashes, melons, potatoes, etc., to reinforce the produce of the many wild food-furnishing plants, nuts, beans, berries, yucca and cactus fruits, grass and sunflower seeds, etc., and the flesh of animals, deer, rabbits, lizards, etc. The cañon Indians I have met here seem to be living much as did their ancestors, though not now driven into rock dens. They are able, erect men, with commanding eyes, which nothing that they wish to see can escape. They are never in a hurry, have a strikingly measured, deliberate, bearish manner of moving the limbs and turning the head, are capable of enduring weather, thirst, hunger, and over-abundance, and are blessed with stomachs which triumph over everything the wilderness may offer. Evidently their lives are not bitter.

The largest of the cañon animals one is likely to see is the wild sheep, or Rocky Mountain bighorn, a most admirable beast, with limbs that never fail, at home on the most nerve-trying precipices, acquainted with all the springs and passes and broken-down jumpable places in the sheer ribbon cliffs, bounding from crag to crag in easy grace and confidence of strength, his great horns held high above his shoulders, wild red blood beating and hissing through every fiber of him like the wind through a quivering mountain pine.

Deer also are occasionally met in the cañon, making their way to the river when the wells of the plateau are dry. Along the short spring streams beavers are still busy, as is shown by the cotton-wood and willow timber they have cut and peeled, found in all the river drift-heaps. In the most barren cliffs and gulches there dwell a multitude of lesser animals, well-dressed, clear-eyed, happy little beasts—wood-rats, kangaroo-rats, gophers, wood-mice, skunks, rabbits, bob cats, and many others, gathering food, or dozing in their sun-warmed dens. Lizards, too, of every kind and color are here enjoying life on the hot cliffs, and making the brightest of them brighter.

Nor is there any lack of feathered people. The golden eagle may be seen, and the osprey, hawks, jays, humming-birds, the mourning-dove, and cheery familiar singers—the black-headed grosbeak, robin, bluebird, Townsend's thrush, and many war-blers, sailing the sky and enlivening the rocks and bushes through all the cañon wilderness.

Here at Hance's river-camp or a few miles above it brave Powell and his brave men passed their first night in the cañon on their adventurous voyage of discovery thirty-three years ago. They faced a thousand dangers, open or hidden, now in their boats gladly sliding down swift, smooth reaches, now rolled over and over in backcombing surges of rough, roaring cataracts, sucked under in eddies, swimming like beavers, tossed and beaten like castaway drift—stout-hearted, undaunted, doing their work through it all. After a month of this they floated smoothly out of the dark, gloomy, roaring abyss into light and safety two hundred miles below. As the flood rushes past us, heavy-laden with desert mud, we naturally think of its sources, its countless silvery branches outspread on thousands of snowy mountains along the crest of the continent, and the life of them, the beauty of them, their history and romance. Its topmost springs are far north and east in Wyoming and Colorado, on the snowy Wind River, Front, Park, and Sawatch ranges, dividing the two ocean waters, and the Elk, Wasatch, Uinta, and innumerable spurs streaked with streams, made famous by early explorers and hunters. It is a river of rivers—the Du Chesne, San Rafael, Yampa, Dolores, Gunnison, Cotchetopa, Uncompahgre, Eagle, and Roaring rivers, the Green and the Grand, and scores of others with branches innumerable, as mad and glad a band as ever sang on mountains, descending in glory of foam and spray from snow-banks and glaciers through their rocky moraine-dammed, beaver-dammed channels. Then, all emerging from dark balsam and pine woods and coming together, they meander through wide, sunny park valleys, and at length enter the great plateau and flow in deep cañons, the beginning of the system culminating in this grand cañon of cañons.

Our warm cañon camp is also a good place to give a thought to the glaciers which still exist at the heads of the highest tributaries. Some of them are of considerable size, especially those on the Wind River and Sawatch ranges in Wyoming and Colorado. They are remnants of a vast system of glaciers which recently covered the upper part of the Colorado basin, sculptured its peaks, ridges, and valleys to their present forms, and extended far out over the plateau region—how far I cannot now say. It appears, therefore, that, however old the main trunk of the Colorado may be, all its wide-spread upper branches and the landscapes they flow through are newborn, scarce at all changed as yet in any important feature since they first came to light at the close of the glacial period.

The so-called Grand Colorado Plateau, of which the Grand Cañon is only one of its well-proportioned features, extends with a breadth of hundreds of miles from the flanks of the Wasatch and Park Mountains to the south of the San Francisco Peaks. Immediately to the north of the deepest part of the cañon it rises in a series of subordinate plateaus, diversified with green meadows, marshes, bogs, ponds, forests, and grovy park valleys, a favorite Indian hunting-ground, inhabited by elk, deer, beaver, etc. But far the greater part of the plateau is good sound desert, rocky, sandy, or fluffy with loose ashes or dust, dissected in some places into a labyrinth of stream-channel chasms like cracks in a dry clay-bed, or the narrow silt crevasses of glaciers—blackened with lava flows, dotted with volcanoes and beautiful buttes, and lined with long continuous escarpments—a vast bed of sediments of an ancient sea-bottom, still nearly as level as when first laid down after being heaved into the sky a mile or two high.

Walking quietly about in the alleys and byways of the Grand Cañon City, we learn something of the way it was made; and all must admire effects so great from means apparently so simple: rain striking light hammer-blows or heavier in streams, with many rest Sundays; soft air and light, gentle sappers and miners, toiling forever; the big river sawing the plateau asunder, carrying away the eroded and ground waste, and exposing the edges of the strata to the weather; rain torrents sawing cross-streets and alleys, exposing the strata in the same way in hundreds of sections, the softer, less resisting beds weathering and receding faster, thus undermining the harder beds, which fall, not only in small weathered particles, but in heavy sheer-cleaving masses, assisted down from time to time by kindly earthquakes, rain torrents rushing the fallen material to the river, keeping the wall rocks constantly exposed. Thus the cañon grows wider and deeper. So also do the side-cañons and amphitheaters, while secondary gorges and cirques gradually isolate masses of the promontories, forming new buildings, all of which are being weathered and pulled and shaken down while being built, showing destruction and creation as one. We see the proudest temples and palaces in stateliest attitudes, wearing their sheets of detritus as royal robes, shedding off showers of red and yellow stones like trees in autumn shedding their leaves, going to dust like beautiful days to night, proclaiming as with the tongues of angels the natural beauty of death.

Every building is seen to be a remnant of once continuous beds of sediments—sand and slime on the floor of an ancient sea, and filled with the remains of animals, and that every particle of the sandstones and limestones of these wonderful structures was derived from other landscapes, weathered and rolled and ground in the storms and streams of other ages. And when we examine the escarpments, hills, buttes, and other monumental masses of the plateau on either side of the cañon, we discover that an amount of material has been carried off in the general denudation of the region compared with which even that carried away in the making of the Grand Cañon is as nothing. Thus each wonder in sight becomes a window through which other wonders come to view. In no other part of this continent are the wonders of geology, the records of the world's auld lang syne, more widely opened, or displayed in higher piles. The whole cañon is a mine of fossils, in which five thousand feet of horizontal strata are exposed in regular succession over more than a thousand square miles of wall-space, and on the adjacent plateau region there is another series of beds twice as thick, forming a grand geological library—a collection of stone books covering thousands of miles of shelving tier on tier conveniently arranged for the student. And with what wonderful scriptures are their pages filled—myriad forms of successive floras and faunas, lavishly illustrated with colored drawings, carrying us back into the midst of the life of a past infinitely remote. And as we go on and on, studying this old, old life in the light of the life beating warmly about us, we enrich and lengthen our own.

# INTERLUDE
## A Grand Canyon Portfolio

# 1898
# The Land of Patience

## David Starr Jordan

David Starr Jordan was one of America's most distinguished
naturalists. Starting in the 1870s and continuing almost until
his death in 1931, he produced more than 1,000 titles—books,
monographs, government reports—on a wide variety of subjects, both within and outside the field of natural history. He
was actively devoted to many good causes, including inter-
national arbitration, and served as the first president of
Stanford University, a position he held at the time of his visit
to the Grand Canyon.

### The Old Repose

NOT its grandeur and beauty, its weird magnificence, its
sublime supremacy; all the world knows this. But it impressed me not the less through its infinite laziness.

While the rest of the earth's crust has been making history and
scenery with all the great earth-molding forces steadily at work,
this corner of the world for ten thousand centuries and more has
rested in the sun.

While mountains were folding and continents taking form, this land of patience lay beneath a warm and shallow sea, the extension of the present Gulf of California. For centuries untold its sands piled up layer on layer.

When at last the uplift of the Sierras changed the sands to dry land, then the forces of erosion began and the sands were torn away as sleepily as they had been deposited before. A mile or two in vertical depth had been stripped away from the whole surface, leaving only flat-topped buttes here and there to testify to the depth of the ancient strata. The flinty limestones half-way down interposed their resistance. The swift river from the glacial mountains which had done this work narrowed its bounds and applied itself more strictly to its business. Cutting at last through the flinty stone, it made quick work of the shales beneath it, and dropping swiftly from level to level, it is now at work on the granite core of the earth at the bottom.

"Running a rapid." Woodcut from *Century Illustrated Monthly*, February 1875

## The River Worked Alone

Even in this it has made fair progress, but the river has done all this alone.

No ice, nor frost, nor earthquake, nor volcanic force has left its mark on the canyon.

Ice would have made a lake of it. Frosts would have changed its cliffs to slopes. Earthquakes would have crumbled its walls, and volcanoes would have smeared them with lava. But none of these forces came to mar or help.

In the simplest, easiest and laziest fashion rocks were deposited in the first place. In the simplest, easiest and laziest fashion they have been torn up again, and a view from the canyon rim almost anywhere shows at a glance how it was all done.

INTERLUDE

# 1902
# The Grand Canyon at Night

## Hamlin Garland

A great many of the writers who have accepted the challenge
of describing the canyon become just as involved in light as in
rock. So much of its appearance depends on time of day,
weather, and season that one comes away from some descriptions quite unsure if there is a *real* canyon; the impressions are
almost too diverse to believe. Some writers were quite out-
spoken on the matter. William Allen White, for example, believed that the "best possible view" of the canyon was from the
rim, and that "sunrise in the canyon is not important."[1] In the
following sketch, Hamlin Garland, Pulitzer Prize-winning
novelist, contrasts his impressions of two remarkably different
Grand Canyons.[2]

## A Twilight Scene

TO me the Grand Canyon possesses two distinct entities. It is
at once a majestic rift in the earth and two ranges of mountains. As I look back upon it, two views of it dominate all others.

The first is from the bank of the turbulent river at twilight,
where I sat alone watching the sun set over the western range,

98

while a superb September moon rose solemnly over the peaks to the east. It is worth while to spend a night alone among these prodigious peaks and listen to the voice of the Colorado as it roars with ever-increasing power, like some imperious nocturnal animal—a dragon with a lion's throat. As the shadows deepen in the lower deeps, beginning to wash like the flood of a spectral purple sea the gray-green mesas of the lower levels, then the river's voice swells till it seems to fill the whole enormous canyon—savage, solemn and persistent.

It was deep night where I sat, while yet on the eastern peaks, a mile above my head, the sun's rays lay in hot red gold. It was instructive to me to see how, one by one, assertive lesser heights sank into shadow, till at last only one or two remained to wear crowns, their lonely grandeur no longer in dispute. And then the moon began to grow great, like the river's voice, pouring among the crags a mystical radiance. As I stood looking at the ragged edge of a cliff set against the great yellow brim, I became aware of something white and mysterious at my right hand; some strange, ghostly, awesome thing had crept upon me silently and was about to envelop me. For an instant my blood thickened with fear. Was it some ghost of the river's dark caverns? It seemed so close I had but to reach out my hand and feel its chill. Each moment it expanded, towering over me. The river seemed suddenly more fiercely menacing of roar, the darkness about my feet deeper, and mysterious rustlings arose in the mesquite. With an effort I whirled and fronted the mysterious presence. It was the face of a cliff across the river, smitten into white radiance and brought near by the marvelous light of the moon. The wall of rock was a half mile away and three thousand feet in height.

## Another Phase

When I came out of the canyon, next day at noon, a sounding wind was blowing and the air had a touch of autumn bitterness in it. The whole vast chasm was roofed with masses of gray clouds ranked closely and hurrying swiftly. The gloom of their presence was magnificent. All the distracting lines of the canyon walls were lost; all tearing, torturing angles softened and made plastic. The haphazard coloring was unified and made harmonious by deep blue curtains of mist. Here, too, was a new phase of the canyon. I began to understand that it had a thousand differing moods, and that no one can know it for what it is who has not lived with it every day of the year. It is like a mountain range—a cloud to-day, a wall of marble to-morrow. When the light falls into it, harsh, direct and searching, it is great, but not beautiful. The lines are chaotic, disturbing—but wait! The clouds and the sunset, the moonrise and the storm will transform it into a splendor no mountain range can surpass. Peaks will shift and glow, walls darken, crags take fire, and gray-green mesas, dimly seen, take on the gleam of opalescent lakes of mountain water. The traveler who goes out to the edge and peers into the great abyss sees but one phase out of hundreds. If he is fortunate it may be one of its most beautiful combinations of color and shadow. But to know it, to feel its majesty, one should camp in the bottom and watch the sunset and the moonrise while the river marches from its lair like an angry lion.

INTERLUDE

# 1903

# Remarks to the
# People of Arizona

## Theodore Roosevelt

Theodore Roosevelt's contributions to the cause of conservation are legendary. More than anyone else in his time, he pro-
moted and fostered intelligent use of the nation's natural resources, and it was the Grand Canyon that moved him to one
of his most daring executive actions in behalf of conservation.
In 1908 he declared the Grand Canyon a national monument,
applying a very liberal interpretation to the Antiquities Act
which Congress had passed in 1906.[1] Roosevelt was challenged,
but his action was eventually upheld by the Supreme Court.

The speech that follows was delivered at the Canyon in 1903,
when he was on an extended western tour, and it has been one
of the most quoted of his conservation statements. In it he expressed the most heartfelt wish of countless canyon lovers and
desert rats, then and now.

I HAVE come here to see the Grand Cañon of Arizona,
because in that cañon Arizona has a natural wonder, which,
so far as I know, is in its kind absolutely unparalleled throughout
the rest of the world. I shall not attempt to describe it, because I
cannot. I could not choose words that would convey or that could

convey to any outsider what that cañon is. I want to ask you to do one thing in connection with it in your own interest, and in the interest of the country.

Keep this great wonder of nature as it now is.

I was delighted to learn of the wisdom of the Santa Fe Railroad in deciding not to build their hotel on the brink of the cañon. I hope you will not have a building of any kind, not a summer cottage, a hotel or anything else, to mar the wonderful grandeur, sublimity, the great loneliness and beauty of the cañon.

Leave it as it is. You cannot improve on it; not a bit. The ages have been at work on it, and man can only mar it. What you can do is to keep it for your children, your children's children and for all who come after you, as one of the great sights which every American, if he can travel at all, should see. Keep the Grand Cañon as it is.

Theodore Roosevelt at Grand Canyon. *Courtesy of the Grand Canyon Natural History Association*

INTERLUDE
# 1914
# The Colorado River

### Owen Wister

Owen Wister wandered the western United States for many
years, drawing from its raw characters and immense proportions the substance of his many stories and novels. He is
best remembered for *The Virginian* (1902), the archetype of the
modern western drama and one of the most popular of all
western novels. His interests in the west were broader than its
people, however, and he did some serious adventuring of his
own, hunting, climbing, and generally bumming around with
friends in search of inspiration and fun (but not necessarily in
that order). The country, its grandness, and even its seeming
inhospitability, attracted and changed Wister, as it changed
other easterners, so that when he returned to the east he would
always have a western perspective to create from. It is hard to
imagine that he could dig much more deeply into his affection
and awe for the west than he does in the following brief
passage.

P ERHAPS this planet does somewhere else contain a thing
like the Colorado River—but that is no matter; we at any

Reprinted with permission of the Macmillan Co., Inc. from Owen Wister's preface to
*Through the Grand Canyon from Wyoming to Mexico* by E. L. Kolb. Copyright 1914 by
Macmillan Publishing Co., Inc., renewed 1942 by Ellsworth L. Kolb.

rate in our continent possess one of nature's very vastest works. After The River and its tributaries have done with all sight of the upper world, have left behind the bordering plains and streamed through the various gashes which their floods have sliced in the mountains that once stopped their way, then the culminating wonder begins. The River has been flowing through the loneliest part which remains to us of that large space once denominated "The Great American Desert" by the vague maps in our old geographies. It has passed through regions of emptiness still as wild as they were before Columbus came; where not only no man lives now nor any mark is found of those forgotten men of the cliffs, but the very surface of the earth itself looks monstrous and extinct. Through such a country as this, scarcely belonging to our era any more than the mammoth or the pterodactyl, scarcely belonging to time at all, does the Colorado approach and enter its culminating marvel. Then, for 283 miles it inhabits a nether world of its own. The few that have ventured through these places and lived are a handful to those who went in and were never seen again. The white bones of some have been found on the shores; but most were drowned; and in this water no bodies ever rise, because the thick sand that its torrent churns along clogs and sinks them.

This place exerts a magnetic spell. The sky is there above it, but not of it. Its being is apart; its climate; its light; its own. The beams of the sun come into it like visitors. Its own winds blow through it, not those of outside, where we live. The River streams down its mysterious reaches, hurrying ceaselessly; sometimes a smooth sliding lap, sometimes a falling, broken wilderness of billows and whirlpools. Above stand its walls, rising through space upon space of silence. They glow, they gloom, they shine. Bend after bend they reveal themselves, endlessly new in endlessly changing veils of colour. A swimming and jewelled blue predominates, as of sapphires being melted and spun into skeins of shifting cobweb. Bend after bend this trance of beauty and awe goes on, terrible as the Day of Judgment, sublime as the Psalms of David. Five thousand feet below the opens and barrens of Arizona, this canyon seems like an avenue conducting to the secret of the universe and the presence of the gods.

# 1922

# An Appreciation of Grand Canyon

## Zane Grey

Zane Grey is best known today as the fabulously prolific author of western novels, many of which are still in print. It is not as widely known that his own life was scarcely less thrilling than his cowboy tales. Made financially independent by his books (at his peak he could turn out one a month), he traveled the world seeking adventure. He was a fanatic fisherman, setting many big-game fishing records, and he hunted many kinds of game, including the lions of the Grand Canyon country. He based novels on the canyon area, and was less a tourist there than most of the other authors in this book. In the Grand Canyon he saw the wildness he liked to portray in his western characters, whose animal nature often seemed so close to the surface:

Once more the strange, infinite silence enfolded the canyon. The far-off golden walls glistened in the sun; farther down, the purple clefts smoked. The many-hued peaks and mesas, aloof from each other, rose out of the depths. It was a grand and gloomy scene of ruin where every glistening descent of rock was but a page of the earth's history.

It brought to my mind a faint appreciation of what time really meant; it spoke of an age of former men; it showed me the lonesome crags of eagles, and the cliff lairs of lions; and it taught mutely, eloquently, a lesson of life—that men are still savages, still driven by a spirit to roam, to hunt, and to slay.[1]

T O see the Grand Canyon full of purple smoke at dawn or sublimely fired at sunset, is to be elevated in soul. To see the red rocks; the alkali flats like snow; the sand dunes so graceful and curved; the long cedar slopes, speckled green and gray, leading up to the bold peaks; the vast black belts of timber; the Navajo facing the sunrise with his silent prayer, the Hopi in his alfalfa fields, or the Apache along the historical Apache trail; the coyote sneaking through the arroyos; the lonely cliff dwellings with their monuments of a vanished race; the endless slopes of sage, green and gray and purple on the heights; the natural stone bridges and the petrified forests—and a thousand more beautiful sights—that is to see Arizona and New Mexico.

The smell of cedar smoke, like burning leaves in autumn; the smell of the desert, dry and clean and somehow new; the smell of the sand and dust, especially after a rain; the tandy odor of the great plateaus of cedar and juniper when your nostrils seem glued as with pitch; and the sweet fragrance of the pine forests, and the indescribable and exhilarating perfume of the purple sage—to know these is to learn the purity of atmosphere never breathed in populous places. To feel the wind in your face; to ride in the teeth of sand storm and flying dust and furious squall; to feel the cold of dawn nip your ears and the heat of noon burn your back; to hear the thunder of the Colorado and the roar of mountain streams, and the rustle of sand through the sage, and the moan of the night breeze in the spruce, the mourn of the wolf and the whistle of the stag; to feel the silence and loneliness of the desert—all this is to grow young again. And to taste the air, water, and meat of the open is to go back hundreds of years, when man was savage and free.

# 1934
# Grand Canyon

John Boynton Priestley

The impressions of a stranger, if not always the most grati-
fying, are often the most enlightening. J. B. Priestley, an Eng-
lishman, became an *afficionado* of the Southwest in the 1930s,
and brought to its description the freshness of the foreign perspective. Among other things, he wondered at American national pride, which seemed to dote on lesser things than this
Grand Canyon. The following pages are taken from a long
article by Priestley which appeared in *Harper's* in 1935, and
they begin as his train approaches the Grand Canyon.

THERE followed a night that had seemed very convenient
and comfortable in the railway time-table, but that
actually was a horrible affair. Just as you became accustomed to
being shunted and were falling asleep, the train stopped and you
were left in an uneasy silence and stillness that made sleep impossible for another hour or two. In the end I must have slept a little,
for I remember waking to find that we were somewhere very high
and that it was snowing. This was Arizona snow. The little station

Reprinted by permission of A. D. Peters & Co. Ltd.

The first white resident of the Grand Canyon, John Hance. Hance entertained tourists for many years with his delicious meals and tall tales of Canyon lore. *Courtesy of the Grand Canyon Natural History Association*

looked as dreary as any I have ever seen. Not so the good-looking young man waiting with the automobile. His costume suggested that a musical comedy was about to begin in these altitudes. He wore a ten-gallon hat and an embroidered cowboy coat, and English riding breeches and long boots, thus making the best of two worlds. The automobile turned a couple of corners and deposited us at a hotel that was almost theatrically Western. It took away our breath, not because it was very beautiful but because we were not used to over-heated rooms at an altitude of about seven thousand feet. But we breakfasted well. We decided then that we had not come all this way to be defeated by a morning's snow, and so agreed to stay at this hotel until the weather cleared and we had seen all we wanted to see. After panting up and down stairs to and from our bedrooms, we ventured out in the mist and sleet.

A few paces in front of the hotel there was nothing: the world stopped; it was after all a flat world and here was the edge. We stared fearfully into the blankness, and after a moment or two there was a swirling, a lifting. Then what breath we had left was clean gone. We were looking into the Grand Canyon.

We were fortunate at the Grand Canyon. There was hardly any kind of weather that did not visit us during our short stay there, so that we saw the country in many different lights. We saw snow falling into the vast gulf, saw clouds stream below us, saw Nineveh and Thebes, rusty in the sunlight, emerge from the mists, saw rainbows arching over the Painted Desert. There is of course no sense at all in trying to describe the Grand Canyon. Those who have not seen it will not believe any possible description. Those who have seen it know that it cannot be described. It passes for a show place and, unlike nearly all other show places in this world, it is far more imposing in reality than in imagination and anticipation. I hear rumors of visitors who were disappointed. The same people will be disappointed at the Day of Judgment. In fact, the Grand Canyon is a sort of landscape Day of Judgment. It is not a show place, a beauty spot, but a revelation. The Colorado River made it; but you feel when you are there that God gave the Colorado River its instructions. The thing is Beethoven's Ninth Symphony in stone and magic light. Even to remember that it is there lifts up the heart. Every member of the Federal Government ought to remind himself, with triumphant pride, that he is on the staff of the Grand Canyon. What a possession for a country! And let me add, how well the country looks after it. The American does not boast enough about his National Parks. Their very existence is something to boast about. The finest pieces of landscape in North America, perhaps in the world, belong to the People and are theirs to enjoy. I take this to be something new in history. It marks a notable advance in civilization. Moreover, the People, through their Federal officers, run these Parks perfectly, and are ideal hosts. All this must not be taken for granted. It is too important. Here is communal ownership working beautifully. Why not turn Chicago into a National Park?

As I stared, hour after hour, at this incredible pageantry of sunlight and chasm, I thought a good deal about America. What do we in Europe see in our minds when we say "America"? We entertain, I think, a confused and not pleasing vision of sky-scrapers, gangsters, tough blondes throwing their legs about, football crowds shouting in chorus, mass-production factories, police automobiles screaming down dark streets, Broadway, Hollywood. What we do not see are the National Parks or the wide ranges of country that flank those Parks: enormous Plains with mountains blue in the distance; a people still busy settling great territories; dams across colossal rivers; roads that pierce deserts and mountains; a land of simple-minded but still heroic engineers, whose fathers saw the Indians retreat. I could not help feeling that the Smart Alecky quality, so prevalent in much-admired contemporary American writing, badly represented this other and bigger America.

You can make any number of wisecracks about the Grand Canyon for example, but you are obviously not expressing the quality of the Grand Canyon; you leave it as you found it. There must be in the soul of this great country a certain large noble simplicity that is hardly finding any verbal expression at all. The people who feel it cannot find the right words. The smart loquacious people cannot feel it. I thought of the fascination that Paris seems to have for so many clever young Americans. Hanging there, wondering, on the brink of the Canyon, this fascination seemed the most preposterous thing and Paris itself a mere distant doll town.

# CHAPTER SIX

# 1909
# Doing the Grand Canyon

## John McCutcheon

John McCutcheon's cartoons delighted generations of Americans. His career, vividly recounted in his autobiography *Drawn From Memory,* took him around the world, in war and peace, from African safari to the Gobi Desert. He counted among his friends some of the leading Americans of the first half of this century.

"Doing the Grand Canyon" was published in 1909 by Fred Harvey, still an important concessioner in the Grand Canyon. At that time, park literature, both educational and instructional, was more often than not produced by private enterprise. McCutcheon's article is amusing enough to deserve reprinting here, but it is also distinguished in another way. That a park concessioner published such a non-commercial booklet, at a time when practically everything printed by many parks was heavily promotional (with maps, rates, and train schedules attached) shows an unusual restraint.

I N describing the Grand Canyon, one should go into a course of literary training and gradually work up to it. He should start off on the Bay of Naples, do that until he has perfected it, then tackle the sunset on the domes and minarets of Stamboul and work on that until he can do it in bogie. Then sunrise on Mount Rigi, the Vale of Cashmir, and other star attractions of nature. Perhaps by this method he might be able to make a try at the Canyon. The great climbers do not begin by ascending a Matterhorn or an Aconcagua the first thing. They do some foothill work first and then by steadily increasing the magnitude of the climb finally are able to negotiate the great peaks. Actors go through years of preparation before they reach their goal—Hamlet well done. Pianists work for years with their ambitions fastened on Liszt's Rhapsodie Hongroise. Violinists work up to Beethoven's Concerto—and so on. When a writer has tackled everything in the line of fancy descriptive writing, he crowns his life work with a pen picture of the Grand Canyon—called by some: "The Greatest Show on Earth." For descriptions of the Canyon, see other writers.

The casual tourist approaches the Canyon with some dread. He fears that he will be disappointed. Surely nothing in nature can equal the expectations of one who has read what great writers have written about this wonderful place. He also fears that if he is disappointed, it may probably be his own fault rather than the Canyon's. It would hurt his pride to be considered as lacking in capacity to appreciate the great beauties of nature, and so, to play safe, he resolves to do full justice to the occasion if it costs him all the adjectives at his command.

It isn't much trouble to reach the Grand Canyon any more. A Pullman Sleeper takes you up to within a couple of hundred yards, and you are supposed to walk the rest of the way. As the time nears when you must meet the test of seeing the Masterpiece of Nature, you experience a peculiar agitation of expectancy. The last mile of railroad riding gives no warning of what lies only a few rods away. When the train stops you climb a flight that leads to the hotel and purposely avoid glancing over in the direction of the Canyon for fear of getting a premature view which would take away the surprise of the supreme moment. You determine that you shall get all the thrill that is possible in one sudden compact shock.

You register leisurely so that you may compose yourself for the supreme moment when you are to get more sightseeing in one glance than is possible any place else in the world—a hole a mile deep and thirteen miles wide, filled with gigantic mountain peaks painted all the colors of the rainbow and fashioned in such beautiful symmetry as to make them seem like great masterpieces of architecture.

The Hotel El Tovar stands near the rim of the Canyon with a level stretch of a hundred feet lying between it and the very edge. A low parapet marks the edge and a number of benches are ranged along for the silent contemplation of the view. Beyond the wall there is nothing. It is as though the wall marked the end of the world and the beginning of infinity. It is not until the sightseer reaches the edge that the full force of the view strikes him with a shock that makes him gasp. All of his set speeches which he has prepared are forgotten as he stands rooted and trembling before the overwhelming spectacle, afraid to utter the adjectives that seem such meager expressions of his emotions.

Silently he stands, gaping at the frightful immensity of the view, and half shrinking from the dreadful depths that shoot thousands of feet directly downward before him. It is as though the world had suddenly dropped away, leaving one clinging on the very edge, with fascinated eyes fixed on mountains so vast and so unexpected as to seem unreal. The sense of unreality is so strong that one imagines himself standing in the middle of a cyclorama building looking at a painting of highly colored mountains and mysterious gorges, so wonderfully done as to suggest an infinity of space. The silence aids in this delusion, and one half expects to go down some steps out into the noise and reality of a street again.

When you speak it is in the hushed respectful tone you would use at a funeral. Any loud exuberance of speech would be irreverent. You have the same awed feeling, multiplied a thousand times, that one experiences as he leans over the tomb of Napoleon in the great shadowey dome of the Hotel des Invalides.

Along the parapet stand silent figures entranced by the wonder of the scene. On the benches sit other figures, all spellbound and awed into silence by the brooding wonder that lies before them. It is like looking into another world—different from anything you have ever seen before.

When I first saw the Canyon a snowstorm was raging over one portion of it. Blue-black clouds were boiling out of the gorges and giving a weird mystery to the Canyon that was anything but earthly. In a moment brilliant red peaks changed to blue as the shadow of the storm swept over them. Great mountains faded in the mist and a moment later reappeared like domes of a city rising from the sea. Off in another part of the Canyon the evening sun was shining brilliantly and down in a gorge a furious rain storm was raging. Stretched before us were all kinds of weather—snow, rain, and sunshine—reminding one of the old-fashioned steel engravings wherein shafts of sunlight streamed down through great boiling masses of silver-tipped clouds—except that instead of black and white, there were blue and dark purple, orange and rosy tint, and wreaths of fleecy clouds whirling in and out of the silent gorges.

I couldn't help thinking of what the old Spanish explorers thought four hundred years ago when they accidentally stumbled, without a moment's warning, on a scene like this. What a shrugging of shoulders there must have been!

As we sat in the comfortable hotel rotunda that evening, surrounded by everything that goes to make life pleasant and comfortable, there would come moments of silence as though each one was vainly struggling to realize that only a few feet away on the right lay that awful brooding chasm, as deep as the ocean and as profoundly silent as the stars.

The real excitement of a trip to the Canyon lies in the ride down one of the trails to the river, a mile below the rim. Most people go down by the Bright Angel Trail, which leads directly down from the Hotel El Tovar, and on which the round trip may be done in about eight hours. The motive power is mule-back, reenforced by a small switch which seems to have little persuasive effect, but imparts a sportylike jauntiness to the rider.

At nine o'clock the caravan assembles in full view of the hotel, much to the dismay of portly ladies in divided skirts who would naturally prefer a less ostentatious start. A cowboy guide has previously determined the number of passengers that the mules are to carry, and one sturdy animal is provided for each passenger. When the latter marches bravely out of the hotel, garbed in borrowed or extemporized riding outfit and with his trusty camera girded about his shoulders, the cowboy asks him (or her) how much he (or she) weighs, and then allots a mule of proportionate strength. Many a mule has been deceived in the weight of ladies of great

"The start from the hotel."

atmospheric displacement. There is much laughter and some nervousness as the adventurers launch themselves, or are launched, into the saddles and the cowboy guide starts gallantly off, followed by a stately and very deliberate caravan of old ladies, young ladies, old gentlemen, young gentlemen, and occasionally a child. There is much forced gayety, but each one is thinking about perils that lie ahead and reassuring himself with the reflection that no one was ever lost in this daring feat, which he now is committed to. The presence of one old lady will have a wonderful effect in bracing up the courage of the whole party. "If she can do it, why, surely I ought to be able to." A few hundred yards from the hotel the caravan turns in toward the Canyon, and the trusty mules with their precious cargoes begin picking their way down the Bright Angel Trail.

The first six or seven hundred feet of the descent is along a snow-covered icy trail that zigzags down at a dizzy angle. Nervous passengers shut their eyes and trust to the mule, whom they hope is as anxious to get home safely as the rider. Of course, when the mule slips there are anxious moments in which the rider wonders how recently the mule was shod, but the latter does not seem to be at all uneasy about the matter. He picks his way downward with deliberate, business-like certainty. He is probably thinking about something to eat. A short way below the rim occurs the first adventure. The caravan is halted while a young man takes a photograph of the crowd. When you return in the evening finished copies will be ready for you, if you wish to purchase them. Of course everybody buys a copy, for who would not give the required amount to have eternal evidence of his daring Israel-Putnam-like dash down the Grand Canyon?

The photographer is very crafty, for he posts his camera in a position overhead that makes the trail look twice as steep as it really is. And that will please you, for in after years when you tell your friends about the memorable ride, you can show them how steep the trail was, and how daring you must necessarily have been to plunge down those ice-bound ledges. Usually, however, the presence in the photograph of some peaceful old lady detracts much from the heroism and daredevil character of your ride.

Of course there *is* a certain amount of danger in going down the Bright Angel Trail. In places this path clings to the face of some

"As they will describe it back home."

dizzy precipice and winds down zigzag ledges that make the rider instinctively shrink away from the outer edge. If the mule should slip, all would be over. BUT—the mule doesn't slip, consequently there is no real danger. The trail is never as narrow or as steep as you will describe it when you get back home. If it were, no living animal could possibly make the trip safely.

One has many things to think of on the ride down. In the first place, there is the possibility of the mule slipping. That is a thought much patronized by the riders. Then there is a chance of a hundred-ton rock being dislodged some place above and bouncing on your head as it passes skippingly to points below. Then there is the thought of fainting, or of vertigo, and other pleasant things to occupy the time, and last, but not least, the glad thought that no one has ever been killed or seriously hurt on the trail, and that lots of elderly people make the trip without minding it.

In the meantime the guide is answering time-honored questions, such as: "Was anyone ever killed on this trail?" "How often do you shoe your mules?" "Where do we have lunch?" "How high is that cliff?" "What makes the stone so red?" "How old is the Canyon?" "Who discovered it?" and "Isn't it remarkable how much those mountains look like old ruins of castles?"

The guide cheerfully gives the required information, whether he knows it or not. It doesn't much matter, for the questioner has asked another before getting the last one answered.

Thirty-four hundred feet below the rim is a beautiful broad plateau on which is situated the little collection of tent cottages called the Indian Gardens. A good spring, a little patch of cultivated garden land, and a sort of a halfway house where cool drinks may be purchased, constitute the settlement. Many people come down and spend the night in the tents, thereby getting an experience which enables them to say afterwards, "When I was roughing it out in Arizona." A long ride across the plateau leads one to the brink of the granite gorge, within which flows, fifteen hundred feet below, the angry, sullen waters of the Colorado River. At one time this plateau was laid out in town lots, for the mining prospectors had reported valuable iron pyrites, and they thought that a fine mining camp would be built up. But this discovery was not of value, and the dreams of a Canyon metropolis went vanishing. Nowadays there are only a few mining claims in the Canyon, most of which are valueless, but are held in the hope that the railroad company will buy them rather than have the scenery mussed up with holes and dump heaps.

The ride down to the river from the Indian Gardens is thrilling, especially the Devil's Corkscrew. This section of the trail—a six-hundred-foot drop down a terrifying zigzag of trail—is not recommended to people who don't like mountain climbing. The path is so steep that riding is unsafe, and the descent and ascent must be made on foot.

By one o'clock you eat your lunch at the edge of the river, with minds somewhat clouded by the realization that you have to go back every foot of that long trip you have come. You do it, however, and at five o'clock the caravan returns like triumphant explorers to the hotel at the top. You look for an easy chair—soft preferred—and discuss with one another your various heroisms of the day.

Sunset is a widely advertised feature of the Grand Canyon. Every promontory that juts out over the chasm has its group of sun worshipers. Adjectives roll out in endless volume as the sun tints the clouds and peaks the most wondrous hues, and the profound depths of the gorges seem even more profound in their purple shadows. Every time a sightseer says something complimentary, a new peak blushes a rosy red. It is an explosion of color, a scrambled rainbow, a thousand square miles of riotous beauty. A man from Indiana who gazed at the scene in silent admiration for a half hour, shook his head and slowly remarked: "Well, sir, it does seem as though the Creator did it just to show what He could do when He tried." In front of the hotel the parade ground along the parapet is always a favorite spot for those who never tire of drinking in the new emotions that come with each succeeding moment. For the Canyon is never the same. There is always something new to see.

Gradually night closes in, and the scenery lovers return, exalted and tired, to the hotel. An hour later the great dining room is full of busy people, and the large lady who looked so funny in her divided skirts, now appears in a bewitching gown and with a slight impediment in her walk.

As we look around at the brilliant room, with its diners from every country in the world, it is hard to realize that we are in a remote desert country and that within one hundred miles are spots never yet explored by man, as well as scores of mountain peaks never yet scaled by adventurous climbers.

After dinner there is the Hopi House to visit. A native dance is scheduled, and an opportunity is offered to those who wish to invest in Indian relics and works of art. The house itself is built in imitation of a genuine Arizona Indian village—entirely of mud and poles—and full of gaily colored rugs of geometric Indian designs.

But the chief ordeal of the day is yet to come. When you go back to the hotel to smoke a final cigar in comfortable ease, you will observe a scene of frenzied activity. Every table is thronged by busy writers. It is the picture-post-card hour, and people are writing cards to everybody they know. It makes you very ill at ease. The fever is hard to resist, and you feel as though you ought to be at work also. After vainly fighting against it for a while, you give up and join the picture-post-card gang. You buy a dozen because you get them cheaper that way, and then write to your six best friends, and finally finish up the other six by writing to the people who will wonder whose initials are signed to the cards.

By ten o'clock the lounging room is empty, and you go away to dream of frightful falls, of mules leaping down thousand-foot cliffs, and of rocks crashing down upon you, inflicting lasting injury. All through the night you have hairbreadth escapes and claw your bedclothes in impotent frenzy. You die a hundred deaths, but in spite of the great mortality you are ready for a good breakfast in the morning.

# CHAPTER SEVEN

# 1909

# Temples of the Grand Canyon

## John Burroughs

John Burroughs, a gentle nature writer from New York's Cat-skill Mountains, made several trips to the west between 1899 and 1910, each with some trepidation. He was a timid man who characterized himself as a "reluctant traveler," and he always seemed surprised at his fame and the attention he received when he traveled.[1] He usually had little to say to his admirers, on account of his shyness, but his reticence was heightened by the scale of western scenes: both in his personal life and in his writing he was more at home with birds, flowers, and secluded forest corners. "I am almost as local as a turtle, and like to poke about in a narrow field," as he once put it.[2]

As if his own natural timidity and the awesome country were not enough of an obstacle to expression, Burroughs visited the Grand Canyon in the company of John Muir (see chapter 5 for Muir's account), who was quite happy to take up the conversational slack. As one member of the party noticed, "Mr. Muir's forte is in monologue."[3]

Burroughs had little to say about his traveling companions in his report on the canyon. He had traveled with Muir to Alaska some years before, and was no doubt accustomed to the man's style. Burroughs was in his early seventies at the Grand

121

Canyon, and saved his energy for a characteristically devotional account of the trip. As fine as his story is, it isn't hard to imagine he was a little relieved to turn his attention to the canyon wren he encountered on the mule ride. The bird, with its reassuring song, fell more comfortably within his normal scope of appreciation than did its huge home.

I N making the journey to the great Southwest—Colorado, New Mexico, Arizona—if one does not know his geology, he is pretty sure to wish he did, there is so much geology scattered over all these Southwestern landscapes crying aloud to be read. The book of earthly revelation, as shown by the great science, lies wide open in that land, as it does in few other places on the globe. Its leaves fairly flutter in the wind, and the print is so large that he who runs on the California limited may read it. Not being able to read it at all, or not taking any interest in it, is like going to Rome or Egypt or Jerusalem, knowing nothing of the history of those lands.

Erosion, erosion—one sees in the West as never before that the world is shaped by erosion. There are probably few or no landscapes in any part of the country from which thousands of feet of rocky strata have not been removed by the slow action of the rain, the frost, the wind; but on our Atlantic seaboard the evidences of it are not patent. In the East, the earth's wounds are virtually all healed, but in the West they are yet raw and gaping, if not bleeding. Then there is so much color in the Western landscape, so many of the warm tints of life, that this fact seems to emphasize their newness, as if they had not yet had time to pale or fade to an ashen gray, under the effects of time, as have our older formations. Indeed, the rocks of the Southwestern region are like volumes of colored plates: not till the books are opened do we realize the splendor of the hues they hold.

Hence it is that when one reaches the Grand Cañon of the Colorado, if he has kept his eyes and mind open, he is prepared to see striking and unusual things. But he cannot be fully prepared for just what he does see, no matter how many pictures of it he may have seen, or how many descriptions of it he may have read.

A friend of mine who took a lively interest in my Western trip wrote me that he wished he could have been present with his kodak when we first looked upon the Grand Cañon. Did he think he could have gotten a picture of our souls? His camera would have shown him only our silent, motionless forms as we stood transfixed by that first view of the stupendous spectacle. Words do not come readily to one's lips, or gestures to one's body, in the presence of such a scene. One of my companions said that the first thing that came into her mind was the old text, "Be still, and know that I am God." To be still on such an occasion is the easiest thing in the world, and to feel the surge of solemn and reverential emotions is equally easy—is, indeed, almost inevitable. The immensity of the scene, its tranquility, its order, its strange, new beauty, and the monumental character of its many forms—all these tend to beget in the beholder an attitude of silent wonder and solemn admiration. I wished at the moment that we might have been alone with the glorious spectacle, or that we might have hit upon an hour when the public had gone to dinner. The smoking and joking tourists sauntering along in apparent indifference, or sitting with their backs to the great geologic drama, annoyed me. I pity the person who can gaze upon the spectacle unmoved. Some are actually terrified by it. I was told of a strong man, an eminent lawyer from a Western city, who literally fell to the earth at the first view, and could not again be induced to look upon it. I saw a woman prone upon the ground near the brink at Hopi Point, weeping silently and long; but from what she afterward told me I know it was not from terror or sorrow, but from the overpowering gladness of the ineffable beauty and harmony of the scene. It moved her like the grandest music. Her inebriate soul could find relief only in tears.

Harriet Monroe was so wrought up by the first view that she says she had to fight against the desperate temptation to fling herself down into the soft abyss, and thus redeem the affront which the very beating of her heart had offered to the inviolable solitude. Charles Dudley Warner said of it, "I experienced for a moment an indescribable terror of nature, a confusion of mind, a fear to be alone in such a presence."

It is beautiful, oh, how beautiful! but it is a beauty that awakens a feeling of solemnity and awe. We called it the "divine abyss." It seems as much of heaven as of earth. Of the many descriptions of it, none seems adequate. To rave over it, or to pour into it a torrent of superlatives, is of little avail. My companion came nearer the mark when she quietly repeated from Revelation, "And he carried me away in the spirit to a great and high mountain, and shewed me that great city, the holy Jerusalem." It does indeed suggest a far-off, half-sacred antiquity, some greater Jerusalem, Egypt, Babylon, or India. We speak of it as a scene: it is more like a vision, so foreign is it to all other terrestrial spectacles, and so surpassingly beautiful.

To ordinary folk, the spectacle is so extraordinary, so unlike everything one's experience has yielded, and so unlike the results of the usual haphazard working of the blind forces of nature, that I did not wonder when people whom I met on the rim asked me what I supposed did all this. I could even sympathize with the remark of an old woman visitor who is reported to have said that she thought they had built the cañon too near the hotel. The enormous cleavage which the cañon shows, the abrupt drop from the brink of thousands of feet, the sheer faces of perpendicular walls of dizzy height, give at first the impression that it is all the work of some titanic quarryman, who must have removed cubic miles of strata as we remove cubic yards of earth.

Indeed, go out to O'Neil's, or Hopi, Point, and, as you emerge from the woods, you get glimpses of a blue or rose-purple gulf opening before you. The solid ground ceases suddenly, and an aërial perspective, vast and alluring, takes its place; another heaven, counter-sunk in the earth, transfixes you on the brink. "Great God!" I can fancy the first beholder of it saying, "what is this? Do I behold the transfiguration of the earth? Has the solid ground melted into thin air? Is there a firmament below as well as above? Has the earth's veil at last been torn aside, and the red heart of the globe been laid bare?" If this first witness was not at once overcome by the beauty of the earthly revelation before him, or terrified by its strangeness and power, he must have stood long, awed, spellbound, speechless with astonishment, and thrilled with delight. He may have seen vast and glorious prospects from mountain-tops, he may have looked down upon the earth and seen it unroll like a map before him; but he had never before looked *into* the earth as through a mighty window or open door, and beheld depths and gulfs of space, with their atmospheric veils and illusions and vast perspectives, such as he had seen from mountain-summits, but with a wealth of color and a suggestion of architectural and monumental remains, and a strange almost unearthly beauty, such as no mountain-view could ever have afforded him.

Three features of the cañon strike one at once: its unparalleled magnitude, its architectural forms and suggestions, and its opulence of color effects—a chasm nearly a mile deep and from ten to twenty miles wide, in which Niagara would be only as a picture upon your walls, in which the pyramids, seen from the rim, would appear only like large tents, and in which the largest building upon the earth would dwindle to insignificant proportions. There are amphitheaters and mighty aisles eight miles long, three or four miles wide, and three or four thousand feet deep; there are room-like spaces eight hundred feet high; there are well-defined alcoves with openings a mile wide; there are niches six hundred feet high overhung by arched lintels; there are pinnacles and rude statues from one hundred to two hundred feet high. Here I am running at once into references to the architectural features and suggestions of the cañon, which must play a prominent part in all faithful attempts to describe it. There are huge, truncated towers, vast, horizontal moldings; there are the semblance of balustrades on the summit of a noble façade. In one of the immense halls we saw, on an elevated platform, the outlines of three enormous chairs, fifty feet or more high, and behind and above them the suggestion of three more chairs in partial ruin. Indeed, there is an opulence of architectural and monumental forms in this divine abyss such as one has never before dreamed of seeing wrought out by the blind forces of nature. These forces have here fore-shadowed all the noblest architecture of the world. Many of the vast carved and ornamental masses which diversify the cañon have been fitly named temples, as Shiva's temple, a mile high, carved out of the red carboniferous limestone, and remarkably symmetrical in its outlines. Near it is the temple of Isis, the temple of Osiris, the Buddha temple, the Horus temple, and the Pyramid of Cheops. Farther to the east is the Diva temple, the Brahma temple, the temple of Zoroaster, and the tomb of Odin. Indeed, everywhere there are suggestions of temples and tombs, pagodas and pyramids, on a scale that no work of human hands can rival. "The grandest objects," says Captain Dutton, "are merged in a congregation of others equally grand." With the wealth of form goes a wealth of color. Never, I venture to say, were reds and browns and grays and vermilions more appealing to the eye than they are as they softly glow in this great cañon. The color scheme runs from the dark, somber hue of the gneiss at the bottom, up through the yellowish brown of the Cambrian layers, and on up through seven or eight broad bands of varying tints of red and vermilion, to the broad yellowish-gray at the top.

One of the smaller of these many geologic temples is called the temple of Isis. How it seems to be resisting the siege of time, throwing out its salients here and there, and meeting the onset of the foe like a military engineer! It is made up of four stories, and its height is about 2500 feet. The finish at the top is a line of heavy wall probably one hundred feet high. The lines of many of these natural temples or fortresses are still more lengthened and attenuated, appearing like mere skeletons of their former selves. The forms that weather out of the formation above this, the Permian, appear to be more rotund, and tend more to domes and rounded hills.

One's sense of the depths of the cañon is so great that it almost makes one dizzy to see the little birds fly out over it, or plunge down into it. One seemed to fear that they, too, would get dizzy and fall to the bottom. We watched a line of tourists on mules creeping along the trail across the inner plateau, and the unaided eye had trouble to hold them; they looked like little red ants. The eye has more difficulty in estimating objects and distances beneath it than when they are above or on a level with it, because it is so much less familiar with depth than with height or lateral dimensions.

One of the remarkable and unexpected things about the cañon is its look of ordered strength. There is no debris or loose wreckage, no tumbled confusion of fallen rocks, but the symmetry and proportion of a city or a fortress. Nearly all the lines are lines of greatest strength. The prevailing profile line everywhere is this:

The upright lines represent lines of cyclopean masonry, and the slant is the talus that connects them, covered with a short, sage-colored growth of some kind, and as soft to the eye as the turf of our fields.

The simple, strong structural lines assert themselves everywhere, and give that look of repose and security characteristic of the scene. The rocky forces always seem to retreat in good order before the onslaught of time; there is neither rout nor confusion. Everywhere they present a calm, upright front to the foe. And the fallen from their ranks, where are they? A cleaner battle-field between the forces of nature one rarely sees.

The weaker portions are of course constantly giving way. The elements incessantly lay siege to these fortresses and take advantage of every flaw or unguarded point, so that what stands has been seven times, yea, seventy times seven tested, and hence gives the impression of impregnable strength. The angles and curves, the terraces and foundations, seem to be the work of some master engineer, with only here and there a toppling rock, and with no chaos or confusion anywhere. All the litter and rubbish seems to have been cleared up, and the job finished. Indeed, such an effect of finished work in mountain and gorge I have never before seen.

I was puzzled to explain to myself the reason of a certain friendly and familiar look which the great abyss had for me. One sees or feels at a glance that it was not born of the throes and convulsions of nature—of earthquake shock or volcanic explosion. It does not suggest the crush of matter and the wreck of worlds. Clearly it is the work of the more gentle and beneficent forces. This probably accounts for the friendly look. Some of the inner slopes and plateaus seemed like familiar ground to me: I must have played upon them when a school-boy. Bright Angel Creek, for some inexplicable reason, recalled a favorite trout stream of my native hills, and the old Cambrian plateau that edges the inner chasm, as we looked down upon it from nearly four thousand feet above, looked like the brown meadow where we played ball in the old schooldays, friendly, tender, familiar, in its slopes and terraces, in its tints and basking sunshine, but grand and awe-inspiring in its depth, its huge walls, and its terrific precipices.

The geologists are agreed that the cañon is only of yesterday in geologic time—the middle Tertiary—and yet behold the duration of that yesterday as here revealed, probably a million years or more! We can no more form any conception of such time than we can of the size of the sun or of the distance of the fixed stars.

The forces that did all this vast delving and sculpturing—the air, the rains, the frost—are as active now as they ever were; but their activity is a kind of slumbering that rarely makes a sign. Only at long intervals is the silence of any part of the profound abyss broken by the fall of loosened rocks or sliding talus. We ourselves saw where a huge splinter of rock had recently dropped from the face of the cliff. In time these loosened masses disappear, as if they melted like ice.

A city not made with hands, but as surely not eternal in the earth. In our humid and severe Eastern climate, frost and ice and heavy rains working together, all these architectural forms would have crumbled long ago, and fertile fields or hill-slopes would have taken their place. In the older Hawaiian Islands, which probably also date from Tertiary times, the rains have carved enormous cañons and amphitheaters out of the hard volcanic rock, in some places grinding the mountains to such a thin edge that a man may literally sit astride them, each leg pointing into opposite valleys. In the next geologic age, these mountains will have largely disappeared. For the same reason, the Grand Cañon will soon, geologically speaking, become a thing of the past.

It seems to take millions of years to tame a mountain, to curb its rude, savage power, to soften its outlines, and bring fertility out of the elemental crudeness and barrenness. But time and the gentle rains of heaven will do it, as they have done it in the East, and as they are fast doing it in the West.

Our little span of human life is far too narrow for us to be a witness of any of the great earth changes. These changes are so slow, oh, so slow, and human history is so brief! So far as we are concerned, the gods of the earth sit in council behind closed doors. All the profound, formative, world-shaping forces of nature go on in a realm that we can reach only through our imaginations. They so far transcend our human experiences that it requires an act of faith to apprehend them. The repose of the hills and the mountains how profound, yet they may be rising or sinking before our very eyes, and we detect no sign. Only on exceptional occasions, during earthquakes or volcanic eruptions, is their dreamless slumber rudely disturbed.

Geologists tell us that from the great plateau in which the Grand Cañon is cut, layers of rock, many thousands of feet thick, were cut away before the cañon was begun.

Starting from the high plateau of Utah, and going south toward the cañon, we descend a grand geologic stairway every shelf or tread of which consists of different formations fifty or more miles broad, from the Eocene, at an altitude of over ten thousand feet at the start, across the cretaceous, the Jurassic, the Triassic, the Permian, to the Carboniferous, which is the bottom, or landing, of the Grand Cañon plateau at an altitude of about five thousand feet. Each step terminates more or less abruptly, the first by a drop of eight hundred feet, ornamented by rows of square obelisks and pilasters of uniform pattern and dimension, "giving the effect," says Major Dutton, "of a gigantic colonnade from which the entablature has been removed or has fallen in ruins." The next step, or platform, the cretaceous, slopes down gradually or dies out on the step beneath it; the Jurassic, which ends in white sandstone cliffs several hundred feet high; then comes the Triassic, which ends in the famous vermilion cliffs thousands of feet high, most striking in color and in form; then comes the Permian tread, which also ends in striking cliffs, with their own style of color and architecture; and, lastly, the great Carboniferous platform in which the cañon itself is carved. Now, all these various strata above the cañon, making at one time a thickness of over a mile, were worn away in Pliocene times, before the cutting of the Grand Cañon began. Had they remained, and been cut through, we should have had a chasm two miles deep instead of one mile.

Two forces, or kinds of forces, have worked together in excavating the cañon: the river, which is the primary factor, and the meteoric forces, which may be called the secondary, as they follow in the wake of the former. The river starts the gash downward, then the aërial forces begin to eat into the sides.

Geologists account for the enormous lateral erosion of the Grand Cañon on the principle of the want of homogeneity of the rocks, the softer strata, by their more rapid decay, undermining the harder layers and causing them to break off in masses. But one sees no evidence of this unequal hardness of the rocks: the huge walls that bound the cañon are everywhere smooth and vertical, and seem to have been eroded equally. There is no talus of broken and fallen fragments at their base, as one would have expected to find had the process of erosion taken this form. Indeed, the rocks seem to have melted and disappeared as if they had been ice.

There is probably another explanation of what we see here. Apart from the mechanical weathering of the rocks as a result of the arid climate, wherein rapid and often extreme changes of temperature take place, causing the surface of the rocks to flake or scale off, there has doubtless been unusual chemical weathering, and this has been largely brought about by the element of iron that all these rocks possess. Their many brilliant colors are imparted to them by the various compounds of iron which enter into their composition. And iron, though the symbol of hardness and strength, is an element of weakness in rocks, as it causes them to oxidize or disintegrate more rapidly. In the marble cañon, where apparently the rock contains no iron, the lateral erosion has been very little, though the river has cut a trench as deep as it has in other parts of its course.

During those days at the cañon how often I thought of the geology of my native hills amid the Catskills, which show the effects of denudation as much older than that shown here as this is older than the washout in the road by this morning's shower. The old red sandstone in which I hoed corn as a farm-boy dates back to middle Paleozoic time, or to the spring of the great geologic year, while the cañon is of the late autumn. Could my native hills have replied to my mute questionings, they would have said: "We were old, old, and had passed through the cañon stage long before the Grand Cañon was born. We have had all that experience, and have forgotten it ages and ages ago. No vestige of our cañons remain. They have all been worn down and obliterated by the strokes of a hand as gentle as that of a passing cloud. Where they were, are now broad, fertile valleys, with rounded knolls and gentle slopes, and the sound of peaceful husbandry. The great ice sheet rubbed us and plowed us, but our contours were gentle and rounded eons before that event. When the Grand Cañon is as old as we are, all its superb architectural features will have long since disappeared, its gigantic walls will have crumbled, and rolling plains and gentle valleys will have taken its place."

All of which seems quite probable. With time enough, the gentle forces of air and water will surely change the whole aspect of this tremendous chasm.

On the second day we made the descent into the cañon on mule-back. That veteran mountain-climber and glacier-meadow Scotsman John Muir pooh-poohed the scheme. "Go up," he said, "and not down. Climb, climb; do not fancy that you can bestride a

John Burroughs and John Muir, third and fifth in line, on Bright Angel Trail. *Courtesy of the Grand Canyon Natural History Association*

mule and go down into that hole and find the glory that lures you from the top."

But we were not to be dissuaded or ridiculed out of the adventure. There is always satisfaction in going to the bottom of things. Then, we wanted to get on more intimate terms with the great abyss, to wrestle with it, if need be, and to feel its power, as well as

131

to behold it. It is not best always to dwell upon the rim of things or to look down upon them from afar. The summits are good, but the valleys have their charm, also; even the valley of humiliation has its lessons. At any rate, four of us were unanimous in our desire to sound that vast profound on mule-back, trusting that the return trip would satisfy our "climbing" aspirations, as it did. The sarcastic Scotsman, seeing that we were not to be ridiculed out of the adventure, reluctantly consented to be one of the misguided party.

It is quite worth while to go down into the cañon on mule-back, if only to fall in love with a mule, and to learn what a sure-footed, careful, and docile creature, when he is on his good behavior, a mule can be. My mule was named "Johnny," and there was soon a good understanding between us. I quickly learned to turn the whole problem of that perilous descent over to him. He knew how to take the sharp turns and narrow shelves of that steep zigzag much better than I did. I do not fancy that the thought of my safety was Johnny's guiding star; his solicitude struck nearer home than that. There was much ice and snow on the upper part of the trail, and only those slender little legs of Johnny's stood between me and a tumble of two or three thousand feet. How cautiously he felt his way with his round little feet, as, with lowered head, he seemed to be scanning the trail critically! Only when he swung around the sharp elbows of the trail did his fore-feet come near the edge of the brink. Only once or twice at such times, as we hung for a breath above the terrible incline, did I feel a slight shudder. One of my companions, who had never before been upon an animal's back, so fell in love with her "Sandy" that she longed for a trunk big enough in which to take him home with her.

It was more than worth while to make the descent to traverse that Cambrian plateau, which from the rim is seen to flow out from the base of the enormous cliffs to the brink of the inner chasm, looking like some soft lavender-colored carpet or rug. I had never seen the Cambrian rocks, the lowest of the stratified formations, nor set my foot upon Cambrian soil. Hence a new experience was promised me. Rocky layers probably two or three miles thick had been worn away from the old Cambrian foundations, and when I looked down upon that gently undulating plateau, the thought of the eternity of time which it represented tended quite as much to make me dizzy as did the drop of nearly four thousand feet. We found it gravelly and desert-like, covered with cacti, low sage-brush, and other growths. The dim trail led us to its edge, where we could look down into the twelve hundred foot V-shaped gash which the river had cut into the dark, crude-looking archæan rock. How distinctly it looked like a new day in creation where the horizontal, yellowish-gray beds of the Cambrian were laid down upon the dark, amorphous, and twisted older granite! How carefully the level strata had been fitted to the shapeless mass beneath it! It all looked like the work of a master mason; apparently you could put the point of your knife where one formation ended and the other began. The older rock suggested chaos and turmoil, the other suggested order and plan, as if the builder had said, "Now upon this foundation we will build our house." It is an interesting fact, the full geologic significance of which I suppose I do not appreciate, that the different formations are usually marked off from one another in just this sharp way, as if each one was indeed the work of a separate day of creation. Nature appears at long intervals to turn over a new leaf and start a new chapter in her great book. The transition from one geologic age to another appears to be abrupt: new colors, new constituents, new qualities appear in the rocks with a suddenness hard to reconcile with Lyell's doctrine of uniformitarianism, just as new species appear in the life of the globe with an abruptness hard to reconcile with Darwin's slow process of natural selection. Is sudden mutation, after all, the key to all these phenomena?

We ate our lunch on the old Cambrian table, placed there for us so long ago, and gazed down upon the turbulent river hiding and reappearing in its labyrinthian channel so far below us.

It is worth while to make the descent in order to look upon the river which has been the chief quarryman in excavating the cañon, and to find how inadequate it looks for the work ascribed to it. Viewed from where we sat, I judged it to be forty or fifty feet broad, but I was assured that it was between two and three hundred feet. Water and sand are ever symbols of instability and inconstancy, but let them work together, and they saw through mountains, and undermine the foundations of the hills.

It is always worth while to sit or kneel at the feet of grandeur, to look up into the placid faces of the earth gods and feel their power, and the tourist who goes down into the cañon certainly has this privilege. We did not bring back in our hands, or in our hats, as the Scotsman said we could not, the glory that had lured us from the top, but we seemed to have been nearer its sources, and to have brought back a deepened sense of the magnitude of the forms, and of the depth of the chasm which we had heretofore gazed upon from a distance. Also we had plucked the flower safely from the nettle danger, always an exhilarating enterprise.

In climbing back, my eye, now sharpened by my geologic reading, dwelt frequently and long upon the horizon where that cross-bedded carboniferous sandstone joins the carboniferous limestone above it. How much older the sandstone looked! I could not avoid the impression that its surface must have formed a plane of erosion ages and ages before the limestone had been laid down upon it.

We had left plenty of ice and snow at the top, but in the bottom we found the early spring flowers blooming, and a settler at what is called the Indian gardens was planting his garden. Here I heard the song of the cañon wren, a new and very pleasing bird-song to me. I think our dreams were somewhat disturbed that night by the impressions of the day, but our day-dreams since that time have at least been sweeter and more comforting, and I am sure that the remainder of our lives will be the richer for our having seen the Grand Cañon.

# CHAPTER EIGHT

# 1913
# Roughing it Deluxe

Irvin Cobb

The Grand Canyon—indeed, the Southwest in general—"arrived" on the American touring scene rather later than many other western wonders. It became a national monument in 1908, when Yellowstone was already a thirty-six-year-old national park. No sooner had it achieved monument status than efforts to make it a national park were intensified. One of the chief proponents of national park status for the canyon was George Horace Lorimer, Editor of the *Saturday Evening Post.* He regularly extolled the canyon's virtues in the *Post,* and in 1912 he sent the well-known writer-humorist Irvin Cobb to visit the canyon as part of a western tour.

Cobb did the assignment justice, at last clearing up the question of why words fail so many who visit the canyon: God neglected to create a word that would "cover it." His greatest service in this article, though, was not in describing the canyon, or even in his impressions of it. It was in his observations of his fellow tourists. The American tourist, whether abroad or at home, had been the target of humorists at least since Sam Clemens produced *The Innocents Abroad* and taught us to laugh at our curious travel habits. Cobb's article was in the best Twain tradition.

I T IS generally conceded that the Grand Cañon beggars description. I shall therefore endeavor to refrain from doing so. Right here at the outset I realize that this is going to be a considerable contract. Nearly everybody, on taking a first look at the Grand Cañon, comes right out and admits its wonders are absolutely indescribable—and then proceeds to write anywhere from two thousand to fifty thousand words, giving the full details. Speaking personally, I wish to say that I do not know anybody who has yet succeeded in getting away with the job.

In the old days when he was doing the literature for the Barnum show, Tody Hamilton would have made the best candidate I can think of. Remember, don't you, how when Tody started in to write about the elephant quadrille you had to turn over to the next page to find the verb? And almost any one of those young fellows who do advertising folders for the railroads would gladly tackle the assignment; in fact, some of them already have—but not with any tumultuous success.

In the presence of the Grand Cañon, language just simply fails you and all the parts of speech go dead lame. When the Creator made it He failed to make a word to cover it. To that extent the thing is incomplete. But if I ever do run across a person who can put down on paper what the Grand Cañon looks like, that party will be my choice to do the story when the Crack of Doom occurs. I can close my eyes now and see the headlines: Judgment Day a complete success! Replete with incident and abounding in surprises—many wealthy families disappointed—full particulars from our special correspondent on the spot!

Starting out from Chicago we had a full trainload. We came from all over—everywhere: from peaceful New England towns full of elm trees and oldline Republicans; from the Middle States; and from the land of chewing tobacco, prominent Adam's apples and hot biscuits—down where the r is silent, as in No'th Ca'lina. And all of us—Northerners, Southerners, Easterners alike—were actuated by a common purpose—we were going West to see the country and rough it—rough it on overland trains better equipped and more luxurious than any to be found in the East; rough it at ten-dollar-a-day hotels; rough it by touring car over the most magnificent automobile roads to be found on this continent. We were a daring lot and resolute; each and every one of us was brave and blithe to endure the privations that such an expedition must inevitably entail. Let the worst come; we were prepared! If there wasn't any of the hothouse lamb, with imported green peas, left we'd worry along on a little bit of the fresh shad roe, and a few conservatory cucumbers on the side. That's the kind of hardy adventurers we were!

Conspicuous among us was a distinguished surgeon of Chicago; in fact, so distinguished that he has had a very rare and expensive disease named for him, which is as distinguished as a physician ever gets to be in this country. Abroad he would be decorated or knighted. Here we name something painful after him and it seems to fill the bill just as well. This surgeon was very distinguished and also very exclusive. After you scaled down from him, riding in solitary splendor in his drawing room, with kitbags full of symptoms and diagnoses scattered round, we became a mixed tourist outfit. I would not want to say that any of the persons on our train were impossible, because that sounds snobbish; but I will say this—some of them were highly improbable.

There was the bride, who put on her automobile goggles and her automobile veil as soon as we pulled out of the Chicago yards and never took them off again—except possibly when sleeping. I presume she wanted to show the rest of us that she was accustomed to traveling at a high rate of speed. If the bridegroom had only bethought him to carry one of those siren horns under his arm, and had tooted it whenever we went round a curve, the illusion would have been complete.

There was also the middle-aged lady with the camera habit. Any time the train stopped, or any time it behaved as though it thought of stopping, out on the platform would pop this lady, armed with her little accordion-plaited camera, with the lens focused and the little atomizer bulb dangling down, all ready to take a few pictures. She snapshotted watertanks, whistling posts, lunch stands, section houses, grade crossings and holes in the snowshed—also scenery, people and climate. A two-by-four photograph of a mountain that's a mile high must be a most splendid reminder of the beauties of Nature to take home with you from a trip.

137

There was the conversational youth in the Norfolk jacket, who was going out West to fill an important vacancy in a large business house—he told us so himself. It was a good selection, too. If I had a vacancy that I wanted filled in such a way that other people would think the vacancy was still there, this youth would have been my candidate.

And finally there was the horse-doctor from a town somewhere in Indiana, who had the upper berth in Number Ten. It seemed to take a load off his mind, on the second morning out, when he learned that he would not have to spend the day up there, but could come down and mingle with the rest of us on a common footing; but right up to the finish of the journey he was uncertain on one or two other points. Every time a conductor came through—Pullman conductor, train conductor or dining-car conductor—he would hail him and ask him this question: "Do I or do I not have to change at Williams for the Grand Cañon?" The conductor—whichever conductor it was—always said "Yes"—he would have to change at Williams. But he kept asking them—he seemed to regard a conductor as a functionary who would deliberately go out of his way to mislead a passenger in regard to an important matter of this kind. After a while the conductors took to hiding out from him and then he began cross-examining the porters, and the smoking-room attendant, and the baggageman, and the flagmen, and the passengers who got aboard down the line in Colorado and New Mexico.

At breakfast in the dining car you would hear his plaintive, patient voice lifted. "Yes, waiter," he would say; "fry 'em on both sides, please. And say, waiter, do you know for sure whether we change at Williams for the Grand Cañon?" He put a world of entreaty into it; evidently he believed the conspiracy against him was widespread. At Albuquerque I saw him leading off on one side a Pueblo Indian who was peddling bows and arrows, and heard him ask the Indian, as man to man, if he would have to change at Williams for the Grand Cañon.

When he was not worrying about changing at Williams he showed anxiety upon the subject of the proper clothes to be worn while looking at the Grand Cañon. Among others he asked me about it. I could not help him. I had decided to drop in just as I was, and then to be governed by circumstances as they might arise; but he was not organized that way. On the morning of the last day, as we rolled up through the pine barrens of Northern Arizona toward our destination, those of us who had risen early became aware of a terrific struggle going on behind the shrouding draperies of his upper berth. Convulsive spasms agitated the green curtains. Muffled swear words uttered in a low but fervent tone filtered down to us. Every few seconds a leg or an arm or a head, or the butt-end of a suitcase, or the bulge of a valise, would show through the curtains for a moment, only to be abruptly snatched back.

Speculation concerning the causes of these strange manifestations ran—as the novelists say—rife. Some thought that, overcome with disappointment at the discovery that we had changed at Williams in the middle of the night, without his knowing anything about it, he was having a fit all alone up there. Presently the excitement abated; and then, after having first lowered his baggage, our friend descended to the aisle and the mystery was explained. He had solved the question of what to wear while gazing at the Grand Cañon. He was dressed in a new golf suit, complete—from the dinky cap to the Scotch plaid stockings. If ever that man visits Niagara I should dearly love to be on hand to see him when he comes out to view the Falls, wearing his bathing suit!

## The Meal's a Success

Some of us aboard that train did not seem to care deeply for the desert; the cactus possibly disappointed others; and the mesquite failed to give general satisfaction, though at a conservative estimate we passed through nine million miles of it. A few of the delegates from the Eastern seaboard appeared to be irked by the tribal dancing of the Hopi Indians—there was not a turkey-trotter in the bunch. The Indian settlements of Arizona are the only terpsichorean centers in this country to which the Young Turk movement has not penetrated yet. Some objected to the plains because they were so flat and plainlike—and some to the mountains because of their exceeding mountainous aspect; but on one point we all agreed—on the uniform excellence of the dining-car service.

It is a powerfully hard thing for a man to project his personality across the grave. In making their wills and providing for the carrying on of their pet enterprises a number of our richest men have endeavored from time to time to disprove this; but, to date, the percentage of successes has not been large. So far as most of us are concerned the burden of proof shows that in this regard we are one with the famous little dog whose name was Rover—when we die, we die all over! Every big success represents the personality of a living man; rarely ever does it represent the personality of a dead man.

The original Fred Harvey is dead—has been dead, in fact, for several years; but his spirit goes marching on across the southwestern half of this country. Two thousand miles from salt water, the oysters that are served on his dining cars do not seem to be suffering from car-sickness. And you can get a beefsteak measuring eighteen inches from tip to tip. There are spring chickens with the most magnificent bust development I ever saw outside of a burlesque show; and the eggs taste as though they might have originated with a hen instead of a cold-storage vault. If there was only a cabaret show going up and down the middle of the car during meals, even New York passengers would be satisfied with the service, I think.

There is another detail of the Harvey system that makes you wonder. Out on the desert, in a dead-gray expanse of silence and sagebrush, your train halts at a junction point that you never even heard of before. There is not much to be seen—a depot, a 'dobe cabin or so, a few frame shacks, a few natives, a few Indians and a few incurably languid Mexicans—and that is positively all there is except that, right out there in the middle of nowhere, stands a hotel big enough and handsome enough for Chicago or New York, built in the Spanish style, with wide patios and pergolas— where a hundred persons might perg at one time—and gay-striped awnings. It is flanked by flower-beds and refreshingly green strips of lawn, with spouting fountains scattered about.

You go inside to a big, spotlessly bright dining room and get as good a meal as you can get anywhere on earth—and served in as good style too. To the man fresh from the East, such an establishment reminds him vividly of the hurry-up railroad lunch places to which he has been accustomed back home—places where the doughnuts are dornicks and the pickles are fossils; the hard-boiled egg looks as though it had got up out of a sick bed to be there, and on the pallid yellow surface of the official pie a couple of hundred flies are seen enacting Custard's Last Stand. It reminds him of them because it is so different. Between Kansas City and the Coast there are a dozen or more of these hotels scattered along the line.

And so, with real food to stay you and one of Tuskegee's brightest graduates to minister to your wants in the sleeper, you come on the morning of the third day out from Chicago to the Grand Cañon; you take one look at it—and instantly you lose all your former standards of comparison! You stand there gazing down the raw, red gullet of that great, gosh-awful gorge, and you feel your self-importance shriveling up to nothing inside of you. You haven't an adjective left to your back. It makes you realize what the sensations would be of one little microbe lost inside of Barnum's fat lady.

I think my preconceived conception of the Cañon was the same conception most people have before they come to see it for themselves—a straight up-and-down slit in the earth, fabulously steep and fabulously deep; nevertheless merely a slit. It is no such thing!

Imagine, if you can, a monster of a hollow approximately some hundreds of miles long and a mile deep, and anywhere from three to sixteen miles wide, with a mountain range—the most wonderful mountain range in the world—planted in it; so that, viewing the spectacle from above, you get the illusion of being in a stationary airship, anchored up among the clouds; imagine these mountainpeaks—hundreds upon hundreds of them—rising one behind the other, stretching away in endless, serried rank until the eye swims and the mind staggers at the task of trying to count them; imagine them splashed and splattered over with all the earthly colors you ever saw and a lot of unearthly colors you never saw before; imagine them carved and fretted and scrolled into all shapes—tabernacles, pyramids, battleships, obelisks, Moorish palaces—the Moorish suggestion is especially pronounced both in colorings and in shapes—monuments, minarets, temples, turrets, castles, spires, domes, tents, tepees, wigwams, shafts!

Imagine other ravines opening from the main one, all with their mouths in her flanks like so many sucking pigs; for there are hundreds of these lesser cañons, and any one of them would be a marvel were they not dwarfed into relative puniness by the mother of the litter. Imagine walls that rise sheer and awful as the wrath of God, and at their base holes where you might hide all the Seven Wonders of the Olden World and never know they were there—or miss them either. Imagine a trail that winds like a snake and climbs like a goat and soars like a bird, and finally bores like a worm and is gone!

Imagine a great cloud-shadow cruising along from point to point, growing smaller and smaller still, until it seems no more than a shifting purple bruise upon the cheek of a mountain, and then, as you watch it, losing itself in a tiny rift which at that distance looks like a wrinkle in the seamed face of an old squaw, but which is probably a huge gash gored into the solid rock for a thousand feet of depth and more than a thousand feet of width.

Imagine, way down there at the bottom, a stream visible only at certain favored points because of the mighty intervening ribs and chines of rock—a stream that appears to you like a torpidly crawling yellow worm, its wrinkling back spangled with dingy white specks, but which is really a wide, deep, brawling, rushing river—the Colorado—full of torrents and rapids; and those white specks are the tops of enormous rocks in its bed!

Imagine—if it be winter—snowdrifts above, with desert flowers blooming alongside the drifts, and down below great stretches of green verdure; imagine two or three separate snowstorms visibly raging at different points, with clear, bright stretches of distance intervening between them, and nearer maybe a splendid rainbow arching downward into the great void; for these meteorological three-ring circuses are not uncommon at certain seasons!

Imagine all this spread out beneath the unflawed turquoise of the Arizona sky and washed in the molten gold of the Arizona sunshine—and if you imagine hard enough and keep it up long enough you may begin, in the course of eight or ten years, to have a faint, a very faint and shadowey conception of this spot where the shamed scheme of creation is turned upside down and the very heart of the world is laid bare before our eyes! Then go to Arizona and see it all for yourself, and you will realize what an entirely inadequate and deficient thing the human imagination is.

It is customary for the newly arrived visitor to take a ride along the edge of the cañon—the rim-drive it is called—with stops at Hopi Point and Pima Point and Mohave Point, and other points where the views are supposed to be particularly good. To do this you get into a buckboard, drawn by horses and driven by a competent young man in a khaki uniform. Leaving behind you a clutter of hotel buildings and station buildings, bungalows and tents, you go winding away through a Government forest reserve containing much fine standing timber and plenty more that is not so fine, it being mainly stunted piñon and gnarly desert growths.

## A View From Number Seven

Presently the road, which is a fine, wide macadamized road, skirts out of the timber and threads along the cañon until it comes to a rocky flange that juts far over. You climb out there and, instinctively treading lightly on your tiptoes and breathing in syncopated breaths, you steal across the ledge, going slowly and carefully until you pause finally upon the very eyelashes of eternity and look down into that great muffin-mold of a cañon.

You are at the absolute jumping-off place. There is nothing between you and the undertaker except six thousand feet, more or less, of dazzling Arizona climate. Below you, beyond you, stretching both ways from you, lie those buried mountains, the eternal herds of the Lord's cattlefold; there are scars upon their sides, like the marks of a mighty branding iron, and in the distance, viewed through the vaporwaves of melting snow, their sides seem to heave up and down like the flanks of panting cattle. Half a mile under you, straight as a man can spit, are gardens of willows and grasses and flowers, looking like tiny green patches, and the tents of a camp looking like white dots; and there is a plateau down there that appears to be as flat as your hand and is seemingly no larger, but actually is of a size sufficient for the evolutions of a brigade of cavalry.

When you have had your fill of this the guide takes you and leads you—you still stepping lightly to avoid starting anything—to a spot from which he points out to you, riven into the face of a vast perpendicular chasm above a cave like a monstrous door, a tremendous and perfect figure seven—the house number of the Almighty Himself. By this I mean no irreverence. If ever Jehovah chose an earthly abiding-place, surely this place of awful, unutterable majesty would be it. You move a few yards farther along and instantly the seven is gone—the shift of shadow upon the rock

Stagecoaches such as this one, shown here on the rim of the Canyon, were the principal means of getting to the Canyon until the railroad was completed in 1901. *Courtesy of the Grand Canyon Natural History Association*

wall has wiped it out and obliterated it—but you do not mourn the loss, because there are still upward of a million things for you to look at.

And then, if you have timed wisely the hour of your coming, the sun pretty soon goes down; and as it sinks lower and lower out of titanic crannies come the thickening shades, making new plays and tricks of painted colors upon the walls—purples and reds and golds and blues and yellows and browns—and the cañon is filled to its very brim with the silence of the coming night.

You stand there, stricken dumb, with your whole being dwarfed yet transfigured; and in the glory of that moment you can even forget the gabble of the lady tourist alongside of you who, after searching her soul for the right words, comes right out and gives the Grand Cañon her cordial endorsement. She pronounces it to be just perfectly lovely! But I said at the outset I was not going to undertake to describe the Grand Cañon—and I'm not. These few remarks were practically jolted out of me and should not be made to count in the total score.

Having seen the cañon—or a little bit of it—from the top, the next thing to do is to go down into it and view it from the sides and the bottom. Most of the visitors follow the Bright Angel Trail which is handily near by and has an assuring name. There are only two ways to do the inside of the Grand Cañon—afoot and on muleback. The hotel provides the necessary regalia, if you have not come prepared—divided skirts for the women and leggings for the men, a mule apiece and a guide to every party of six or eight.

At the start there is always a lot of nervous chatter—airy persiflage flies to and fro and much laughing is indulged in. But it has a forced, strained sound, that laughter has; it does not come from the heart, the heart being otherwise engaged for the moment. Down a winding footpath moves the procession, with the guide in front, and behind him in single file his string of pilgrims—all as nervous as cats and some holding to their saddle-pommels with deathgrips. Just under the first terrace a halt is made while the official photographer takes a picture; and when you get back he has your finished copy ready for you, so you can see for yourself just how pale and haggard and wall-eyed and like a typhoid patient you looked.

145

The parade moves on. All at once you notice that the person immediately ahead of you has apparently ridden right over the wall of the cañon. A moment ago his arched back loomed before you; now he is utterly gone. It is at this point that some tourists tender their resignations—to take effect immediately. To the credit of the sex be it said, the statistics show that fewer women quit here than men. Nearly always there is some man who remembers where he left his umbrella or something, and he goes back after it and forgets to return.

## Farewell to the World

In our crowd there was one person who left us here. He was a circular person; about forty per cent of him, I should say, rimmed with jelly. He climbed right down off his mule. He said:

"I'm not scared myself, you understand, but I've just recalled that my wife is a nervous woman. She'd have a fit if she knew I was taking this trip! I love my wife, and for her sake I will not go down this cañon, dearly as I would love to." And with that he headed for the hotel. I wanted to go with him. I wanted to go along with him and comfort him, help him have his chill, and if necessary send a telegram for him to his wife—she was in Pittsburg—telling her that all was well. But I did not. I kept on. I have been trying to figure out ever since whether this showed courage on my part or cowardice.

Over the ridge and down the steep declivity beyond goes your mule, slipping a little. He is reared back until his rump almost brushes the trail; he grunts mild protests at every lurching step and grips his shoecalks into the half-frozen path. You reflect that thousands of persons have already done this thing; that thousands of others—men, women and children—are going to do it, and that no serious accident has yet occurred—which is some comfort, but not much. The thought comes to you that, after all, it is a very bright and beautiful world you are leaving behind. You turn your head to give it a long, lingering farewell, and try to put your mind on something cheerful—such as your life insurance. Then something happens.

The trail, that has been slanting at a downward angle which is a trifle steeper than a ship's ladder, but not quite so steep perhaps as a board fence, takes an abrupt turn to the right. You duck your

146

head and go through a little tunnel in the rock, patterned on the same general design of the needle's eye that is going to give so many of our prominent captains of industry trouble in the hereafter. And as you emerge on the lower side you forget all about your life-insurance papers and freeze to your pommel with both hands, and cram your poor cold feet into the stirrups—even in warm weather they'll be good and cold—and all your vital organs come up in your throat, where you can taste them. If anybody had shot me through the middle just about then he would have inflicted only a flesh wound.

You have come out on a place where the trail clings to the sheer side of the dizziest, deepest chasm in the known world. One of your legs is scraping against the everlasting granite; the other is dangling over half a mile of fresh mountain air. The mule's off hind hoof grates and grinds on the flinty trail, dislodging a fair-sized stone that flops over the verge. You try to look down and see where it is going and find you haven't the nerve to do it—but you can hear it falling from one narrow ledge to another, picking up other stones as it goes until there must be a fair-sized little avalanche of them cascading down. The sound of their roaring, racketing passage dies almost out, and then there rises up to you from those unutterable depths a dull, thuddy little sound—those stones have reached the cellar! Then to you there comes the pleasing reflection that if your mule slipped and you fell off and were dashed to fragments, they would not be large, mussy, irregular fragments, but little teeny-weeny fragments, such as would not bring the blush of modesty to the cheek of the most fastidious.

## My Melancholy Mule

Only your mule never slips off! It is contrary to a mule's religion and politics, and all his traditions and precedents, to slip off. He may slide a little and stumble once in a while, and he may, with malice aforethought, try to scrape you off against the outjutting shoulders of the trail; but he positively will not slip off. It is not because he is interested in you. A tourist on the cañon's rim a simple tourist is to him and nothing more; but he has no intention of getting himself hurt. Instinct has taught that mule it would be to him a highly painful experience to fall a couple of thousand feet or so and light on a pile of rocks; and therefore, through motives that are purely selfish, he studiously refrains from so doing. When the Prophet of old wrote, "How beautiful upon the mountains are the feet of him," and so on, I judge he had reference to a mule on a narrow trail.

My mule had one very disconcerting way about him—or, rather, about her, for it was of the gentler sex. When she came to a particularly scary spot, which was every minute or so, she would stop dead still. I concurred in that part of it heartily. But then she would face outward and crane her neck over the fathomless void of that bottomless pit, and for a space of moments would gaze steadily downward, with a suicidal gleam in her eyes. It worried me no little; and if I had known, at the time, that she had a German name it would have worried me even more, I guess. But either the time was not ripe for the rash act or else she abhorred the thought of being found dead in the company of a mere tourist, so she did not leap off into space, but restrained herself; and I was very grateful to her for it. It made a bond of sympathy between us.

On you go, winding on down past the red limestone and the yellow limestone and the blue sandstone, which is green generally; past huge bat caves and the big nests of pack-rats, tucked under shelves of Nature's making; past stratified millions of crumbling seashells that tell to geologists the tale of the salt-water ocean that once on a time, when the world was young and callow, filled this hole brim full; and presently, when you have begun to piece together the tattered fringes of your nerves, you realize that this cañon is even more wonderful when viewed from within than it is when viewed from without. Also, you begin to notice now that it is most extensively autographed.

Apparently about every other person who came this way remarked to himself that this cañon was practically completed and only needed his signature as collaborator to round it out—so then he signed it and it was a finished job. Some of them brought down colored chalk and stencils, and marking pots, and paints and brushes, and cold chisels to work with, which must have been a lot of trouble, but was worth it—it does add so greatly to the beauty of the Grand Cañon to find it spangled over with such names as you could hear paged in almost any dollar-a-day American-plan hotel! The guide pointed out a spot where one of these inspired authors climbed high up the face of a white cliff and, clinging there, carved out in letters a foot long his name; and it was one of those names that, inscribed upon a register, would instinctively cause any room clerk to reach for the key to an inside one, without bath. I regret to state that nothing happened to this person. He got down safe and sound; it was a great pity too.

By the Bright Angel Trail it is three hours on a mule to the plateau, where there are green summery things growing even in midwinter, and where the temperature is almost sultry; and it is an hour or so more to the riverbed, down at the very bottom. When you finally arrive there and look up you do not see how you ever got down, for the trail has magically disappeared; and you feel morally sure you are never going to get back. If your mule were not under you pensively craning his head rearward in an effort to bite your leg off, you would almost be ready to swear the whole thing was an optical illusion or a wondrous dream.

Under these circumstances it is not so strange that some travelers who have been game enough until now suddenly weaken. Their nerves capsize and the grit runs out of them like sand out of an overturned pail.

All over this part of Arizona they tell you the story of the lady from the southern part of the state—she was a school teacher and the story has become an epic—who went down Bright Angel one morning and did not get back until two o'clock the following morning; and then she came against her will in a litter borne by two tired guides, while two others walked beside her and held her hands; and she was protesting at every step that she positively could not and would not go another inch. She was as hysterical as a treeful of poll-parrots; her hat was lost, her glasses were gone, and her hair hung down her back—altogether she was a mournful sight to see.

Likewise the natives will tell you the tale of a man who made the trip by crawling round the more sensational corners upon his hands and knees.

Coming back out of the Grand Cañon is an even more inspiring and amazing performance than going down. But by now—anyhow this was my experience, and they tell me it is the common experience—you are beginning to get used to the sensation of skirting along the raw and ragged verge of nothing. Narrow turns

where, going down, your hair pushed your hat off no longer af-
fright you; you now take them jauntily—almost debonairly. You
feel like an old mountain-scaler, and your soul begins to crave for
a trip with a few more thrills to the square inch in it. You get your
wish. You go down Hermit Trail, the middle name of which is
thrills; and there you make the acquaintance of the Hydrophobic
Skunk. The Hydrophobic Skunk is a creature of such surpassing
accomplishments and vivid personality that I feel he is entitled to
a new chapter.

The Hydrophobic Skunk resides at the extreme bottom of the
Grand Cañon and, next to a Southern Republican who never
asked for a Federal office, is the rarest of living creatures. He is so
rare that nobody ever saw him—that is, nobody except a native. I
met plenty of tourists who had seen people who had seen him, but
never a tourist who had seen him with his own eyes. In addition
to being rare he is highly gifted.

I think almost anybody will agree with me that the common,
ordinary skunk has been most richly dowered by Nature. To
adorn a skunk with any extra qualifications seems as great a waste
of the raw material as painting the lily or gilding refined gold. He
is already amply equipped for outdoor pursuits. Nobody inten-
tionally shoves him round; everybody gives him as much room as
he seems to need. He commands respect—nay, more than that,
respect and veneration—wherever he goes. Joy-riders never run
him down and foot passengers avoid crowding him into a corner.
You would think Nature had done amply well by the skunk; but
no—the Hydrophobic Skunk comes along and upsets all these
calculations. Besides carrying the traveling credentials of an ordinary skunk, he is rabid in the most rabidissimus form. He is not
mad just part of the time, like one's relatives by marriage—and not
mad most of the time, like a railroad ticket agent—but mad all the
time—incurably, enthusiastically and unanimously mad! He is
mad and he is glad of it.

## Over the Fence is Out

We made the acquaintance of the Hydrophobic Skunk when we
rode down Hermit Trail. The casual visitor to the Grand Cañon
first of all takes the rim drive; then he essays Bright Angel Trail,
which is sufficiently scary for his purposes until he gets used to it;

and after that he grows more adventurous and tackles Hermit Trail, which is a marvel of corkscrew convolutions, gimleting its way down this red abdominal gash of a cañon to the very gizzard of the world.

Alongside the Hermit, traveling the Bright Angel is the same as gathering the myrtles with Mary; but the civil engineers who worked out the scheme of the Hermit and made it wide and navigable for ordinary folks were bright young men. They laid a wall along its outer side all the way from the top to the bottom. Now this wall is made of loose stones racked up together without cement, and it is nowhere more than a foot or a foot and a half high. If your mule ever slipped—which he never does—or if you rolled off on your own hook—which has not happened to date—that puny little wall would hardly stop you—might not even cause you to hesitate. But some way, intervening between you and a thousand feet or so of uninterrupted fresh air, it gives a tremendous sense of security. Life is largely a state of mind anyhow, I reckon.

As a necessary preliminary to going down Hermit Trail you take a buckboard ride of ten miles—ten wonderful miles! Almost immediately the road quits the rocky, bare parapet of the gorge and winds off through a noble, big forest that is a part of the Government reserve. Jays that are twice as large and three times as vocal as the Eastern variety weave blue threads in the green background of the pines; and if there is snow upon the ground its billowy white surface is crossed and crisscrossed with the dainty tracks of coyotes, and sometimes with the broad, furry marks of the wildcat's pads. The air is a blessing and the sunshine is a benediction.

Away off yonder, through a break in the conifers, you see one lone and lofty peak with a cap of snow upon its top. The snow fills the deeper ravines that furrow its side downward from the summit so that at this distance it looks as though it were clutched in a vast white owl's claw; and generally there is a wispy cloud caught on it like a white shirt on a poor man's Monday washpole. Or, huddled together in a nest formation like so many speckled eggs, you see the clutch of little mottled mountains for which nobody seems to have a name. If these mountains were in Scotland, Sir Walter Scott and Bobby Burns would have written about

151

them and they would be world-famous, and tourists from America would come and climb their slopes, and stand upon their tops, and sop up romance through all their pores. But being in Arizona, dwarfed by the heaven-reaching ranges and groups that wall them in north, south and west, they have not even a Christian name to answer to.

Anon—that is to say, at the end of those ten miles—you come to the head of Hermit Trail. There you leave your buckboard at a way station and mount your mule. Presently you are crawling downward, like a fly on a board fence, into the depths of the chasm. You pass through rapidly succeeding graduations of geology, verdure, scenery and temperature. You ride past little sunken gardens full of wild flowers and stunty fir trees, like bits of Old Japan; you climb naked red slopes crowned with the tall cactus, like Old Mexico; you skirt bald, bare, blistered vistas of desolation, like Old Perdition. You cross Horsethief's Trail, which was first traced out by the moccasined feet of marauding Apaches and later was used by white outlaws fleeing northward with their stolen pony herds.

## Looking Down at Blythe's Abyss

You pass above the gloomy shadows of Blythe's Abyss and wind beneath a great box-shaped formation of red sandstone set on a spindle rock and balancing there in dizzy space like Mohammed's coffin; and then, at the end of a mile-long jog along a natural terrace stretching itself mid-way between Heaven and the other place, you come to the residence of Shorty, the official hermit of the Grand Cañon.

Shorty is a little, gentle old man, with warped legs and mild blue eyes and a set of whiskers of such indeterminate aspect that you cannot tell at first look whether they are just coming out or just going back in. He belongs—or did belong—to the vast vanishing race of oldtime gold prospectors. Halfway down the trail he does light housekeeping under an accommodating flat ledge that pouts out over the pathway like a snuffdipper's under lip. He has a hole in the rock for his chimney, a breadth of weathered gray canvas for his door and an eight-mile stretch of the most marvelous panorama on earth for his front yard. He minds the trail and watches out for the big boulders that sometimes fall in the night; and, except in the tourist season, he leads a reasonably quiet existence.

Alongside of Shorty, Robinson Crusoe was a tenement-dweller, and Jonah, week-ending in the whale, had a perfectly uproarious time; but Shorty thrives on a solitude that is too vast for imagining. He would not trade jobs with the most potted potentate alive—only sometimes in midsummer he feels the need of a change stealing over him, and then he goes afoot out into the middle of Death Valley and spends a happy vacation of five or six weeks with the Gila monsters and the heat. He takes Toby with him.

Toby is a gentlemanly little woolly dog built close to the earth like a carpet sweeper, with legs patterned crookedly—after the model of his master's. Toby has one settled prejudice: he dislikes Indians. You have only to whisper the word "Injun" and instantly Toby is off, scuttling away to the highest point that is handy. From there he peers all round looking for red invaders. Not finding any he comes slowly back, crushed to the earth with disappointment. Nobody has ever been able to decide what Toby would do with the Indians if he found them; but he and Shorty are in perfect accord. They have been associated together ever since Toby was a pup and Shorty went into the hermit business, and that was nine years ago. Sitting cross-legged on a flat rock like a little gnome, with his puckered eyes squinting off at space, Shorty told us how once upon a time he came near losing Toby.

"Me and Toby," he said, "was over to Flagstaff, and that was several years ago. There was a saloon man over there owned a bulldog and he wanted that his bulldog and Toby should fight. Toby can lick mighty nigh any dog alive; but I didn't want that Toby should fight. But this here saloon man wouldn't listen. He sicked his bulldog on to Toby and in about a minute Toby was taking that bulldog all apart.

"This here saloon man he got mad then—he got awful mad. He wanted to kill Toby and he pulled out his pistol. I begged him mighty hard please not to shoot Toby—I did so! I stood in front of Toby to protect him and I begged that man not to do it. Then some other fellows made him put up his gun, and me and Toby came on away from there." His voice trailed off. "I certainly would 'a' hated to lose Toby. We set a heap of store by one another—don't we, dog?" And Toby testified that it was so—testified with wriggling body and licking tongue and dancing eyes and a madly wagging stump tail.

## Methodical Tom and His Methods

As we mounted and jogged away we looked back, and the pair of them—Shorty and Toby—were sitting there side by side in perfect harmony and perfect content; and I could not help wondering, in a country where we sometimes hang a man for killing a man, what would have been adequate punishment for a brute who would kill Toby and leave Shorty without his partner! In another minute, though, we had rounded a jutting sandstone shoulder and they were out of sight.

About that time Johnny, our guide, felt moved to speech, and we hearkened to his words and hungered for more, for Johnny knows the ranges of the Northwest as a city dweller knows his own little side street. In the fall of the year Johnny comes down to the Cañon and serves as a guide a while; and then, when he gets so he just can't stand associating with tourists any longer, he packs his warbags and journeys back to the Northern Range and enjoys the company of cows a spell. Cows are not exactly exciting, but they don't ask fool questions.

A highly competent young person is Johnny and a cowpuncher of parts. Most of the Cañon guides are cowpunchers—accomplished ones, too, and of high standing in the profession. With a touch of reverence Johnny pointed out to us Sam Scovel, the greatest bronco buster of his time, now engaged in piloting tourists.

"Can he ride?" echoed Johnny in answer to our question. "Scovel could ride an earthquake if she stood still long enough for him to mount! He rode Steamboat—not Young Steamboat, but Old Steamboat! He rode Rocking Chair, and he's the only man that ever did do that and not be called on in a couple of days to attend his own funeral."

This day he told us about one Tom, who lived up in Wyoming, where Johnny came from. It appeared that in an easier day Tom was hired by some cattle men to thin out the sheep herders who insisted upon invading the public ranges. By Johnny's account Tom did the thinning with conscientious attention to detail and gave general satisfaction for a while; but eventually he got careless in his methods and took to killing parties who were under the protection of the game laws. Likewise his own private collection of yearlings began to increase with a rapidity which was only to be accounted for on the theory that a large number of calves were coming into the world with Tom's brand for a birthmark. So he lost popularity. Several times his funeral was privily arranged, but on each occasion was postponed owing to the failure of the corpse to be present. Finally he killed a young boy and was caught and convicted and legally hanged.

"Tom was mighty methodical," said Johnny. "He got five hundred a head for killing sheep herders—that was the regular tariff. Every time he bumped one off he'd put a stone under his head, which was his private mark—a kind of a duebill, as you might say. And when they'd find that dead herder with the rock under his head they'd know there was another five hundred comin' to Tom on the books; they always paid it too. Once in a while, though, he'd cut loose in a saloon and garner in some fellows that wasn't sheep herders."

We went on and on at a lazy muletrot, hearing the unwritten annals of the range from one who had seen them enacted at first hand. Pretty soon we passed a herd of burros with mealy, dusty noses and spotty hides, feeding on prickly pears and rock lichens; and just before sunset we slid down the last declivity out upon the plateau and came to a camp as was a camp!

This was roughing it de luxe with a most de-luxey vengeance! Here were three tents, or rather three canvas houses, with wooden half-walls; and they were spick-and-span inside and out, and had glass windows in them and doors and matched wooden floors. The one that was a bedroom had gay Navajo blankets on the floor, and a stove in it, and a little bureau, and a washstand with white towels and good lathery soap. And there were two beds— not cots or bunks, but regular beds—with wire springs and mattresses and white sheets and pillowslips. They were not veteran sheets and vintage pillowslips either, but clean and spotless ones. The mess tent was provided with a table with a clean cloth to go over it, and there were china dishes and china cups and shiny knives, forks and spoons. Every scrap of this equipment had been brought down from the top on burro packs. The Grand Cañon is scenically artistic, but it is a non-producing district. And outside there was a corral for the mules; a canvas storehouse; hitching stakes for the burros; a Dutch oven, and a little forge where the guides sometimes shoe a mule. They aren't blacksmiths; they merely have to be. Bill was in charge of the camp—a dark, rangy, good-looking young leading man of a cowboy, wearing his blue shirt and his red neckerchief with an air. He spoke with the soft Texas drawl and in his way was as competent as Johnny.

## A Moonrise Up to Belasco's Best

The sun, which had been winking farewells to us over the rim above, dropped out of sight as suddenly as though it had fallen into a well. From the bottom the shadows went sliding up the glooming walls of the gorges, swallowing up the yellow patches of sunlight that still lingered near the top like blacksnakes swallowing eggs. Every second the colors shifted and changed; what had been blue a moment before was now purple and in another minute would be a velvety black. A little lost ghost of an echo stole out of a hole, feebly mocking our remarks and making them sound cheap and tawdry.

Then the new moon showed as a silver fish balancing on its tail and arching itself like a hooked skipjack. In a turquoise sky the stars popped out like pinpricks and the peace that passes all understanding came over us. I wish to take advantage of this opportunity to say that, in my opinion, David Belasco has never done anything in the way of scenic effects to beat a moonrise in the Grand Cañon.

I reckon we might have been there until now—my companion and I—soaking our souls in the unutterable beauty of that place, only just about that time we smelled something frying. There was also a most delectable sputtering sound as of fat meat turning over on a hot skillet; but just the smell alone was a square meal for a poor family. The meeting adjourned by acclamation. Just because a man has a soul is no reason he shouldn't have an appetite.

That Johnny certainly could cook! Served on china dishes upon a cloth-covered table, we had mounds of fried steaks and shoals of fried bacon; and a bushel, more or less, of sheepherder potatoes; and green peas and sliced peaches out of cans; and sourdough biscuits as light as kisses and much more filling; and fresh butter and fresh milk; and coffee as black as your hat and strong as sin. How

156

easy it is for civilized man to become primitive and comfortable in his way of eating, especially if he has just ridden ten miles on a buckboard and nine more on a mule and is away down in the bottom of the Grand Cañon—and there is nobody to look on disapprovingly when he takes a bite that would be a credit to a steam shovel!

Despite all reports to the contrary, I wish to state that it is no trouble at all to eat green peas off a knifeblade—you merely mix them in with potatoes for a cement; and fried steak—take it from an old steak-eater—tastes best when eaten with those tools of Nature's own providing, both hands and your teeth. An hour passed—busy, yet pleasant—and we were both gorged to the gills and had reared back with our cigars lit to enjoy a third jorum of black coffee apiece, when Johnny, speaking in an offhand way to Bill, who was still hiding away biscuits inside of himself like a parlor prestidigitator, said:

"Seen any of them old hydrophobies the last day or two?"

"Not so many," said Bill casually. "There was a couple out last night pirootin' round in the moonlight. I reckon, though, there'll be quite a flock of 'em out tonight. A new moon always seems to fetch 'em up from the river."

Both of us quit blowing on our coffee and we put the cups down. I think I was the one who spoke.

"I beg your pardon," I asked, "but what did you say would be out tonight?"

"We were just speakin' to one another about them Hydrophoby Skunks," said Bill apologetically.

## Sociable Little Cusses

I laid down my cigar too. I admit I was interested.

"Oh!" I said softly—like that.

"Yes," said Johnny. "I reckon there's liable to be one come shovin' his old nose into that door any minute. Or probably two—they mostly travels in pairs—sets, as you might say."

"You'd know one the minute you saw him though," said Bill. "They're smaller than a regular skunk and spotted where the other kind is striped. And they got little red eyes. You won't have no trouble at all recognizin' one."

157

It was at this juncture that we both got up and moved back by the stove. It was warmer there and the chill of evening seemed to be settling down noticeably.

"Funny thing about Hydrophoby Skunks," went on Johnny after a moment of pensive thought—"mad, you know!"

"What makes them mad?" The two of us asked the question together.

"Born that way!" explained Bill—"mad from the start, and won't never do nothin' to get shut of it."

"Ahem—they never attack humans, I suppose?"

"Don't they?" said Johnny, as if surprised at such ignorance. "Why, humans is their favorite pastime! Humans is just pie to a Hydrophoby Skunk. It ain't really any fun to be bit by a Hydrophoby Skunk neither." He raised his coffee cup to his lips and imbibed deeply.

"You certainly said something then, Johnny," stated Bill. "You see," he went on, turning to us, "they aim to catch you asleep and they creep up right soft and take holt of you—take holt of a year usually—and clamp their teeth and just hang on for further orders. Some says they hang on till it thunders, same as snappin' turtles. But that's a lie, I judge, because there's weeks on a stretch down here when it don't thunder. All the cases I ever heard of they let go at sun-up."

"It is right painful at the time," said Johnny, taking up the thread of the narrative; "and then in nine days you go mad yourself. Remember that fellow the Hydrophoby Skunk bit down here by the rapids, Bill? Let's see now—what was that man's name?"

"Williams," supplied Bill—"Heck Williams. I saw him at Flagstaff when they took him there to the hospital. That guy certainly did carry on regardless. First he went mad and his eyes turned red, and he got so he didn't have no real use for water—well, them prospectors don't never care much about water anyway—and then he got to snappin' and bitin' and foamin' so's they had to strap him down to his bed. He got loose though."

"Broke loose, I suppose?" I said.

"No, he bit loose," said Bill with the air of one who would not deceive you even in a matter of small details.

"Do you mean to say he bit those leather straps in two?"

"No, sir; he couldn't reach them," explained Bill, "so he bit the bed in two. Not in one bite, of course," he went on. "It took him several. I saw him after he was laid out. He really wasn't no credit to himself as a corpse."

## The Coming of the Little Redeyes

I'm not sure, but I think my companion and I were holding hands by now. Outside we could hear that little lost echo laughing to itself. It was no time to be laughing either. Under certain circumstances I don't know of a lonelier place anywhere on earth than that Grand Cañon.

Presently my friend spoke, and it seemed to me his voice was a mite husky. Well, he had a bad cold.

"You said they mostly attack persons who are sleeping out, didn't you?"

"That's right too," said Johnny, and Bill nodded in affirmation.

"Then, of course, since we sleep indoors everything will be all right," I put in.

"Well, yes and no," answered Johnny. "In the early part of the evening a hydrophoby is liable to do a lot of prowlin' round outdoors; but toward mornin' they like to git into camps—they dig up under the side walls or come up through the floor—and they seem to prefer to get in bed with you. They're cold-blooded, I reckin, same as rattlesnakes. Cool nights always do drive 'em in, seems like."

"It's going to be sort of coolish tonight," said Bill casually.

It certainly was. I don't remember a chillier night in years. My teeth were chattering a little—from cold—before we turned in. I retired with all my clothes on, including my boots and leggings, and I wished I had brought along my earmuffs. I also buttoned my watch into my lefthand shirt pocket, the idea being if for any reason I should conclude to move during the night I should be fully equipped for traveling. The door would not stay closely shut—the doorjamb had sagged a little and the wind kept blowing the door ajar. But after a while we dozed off.

It was one-twenty-seven A.M. when I woke with a violent start. I know this was the exact time because that was when the jar stopped my watch. I peered about me in the darkness. The door was wide open—I could tell that. Down on the floor there was a dragging, scuffling sound, and from almost beneath me a pair of small red eyes peered up phosphorescently.

"He's here!" I said to my companion as I emerged from my blankets; and he, waking instantly, seemed instinctively to know whom I meant. I used to wonder at the ease with which a cockroach can climb a perfectly smooth wall and run across the ceiling. I know now that to do this is the easiest thing in the world—if you have the proper incentive behind you. I had gone up one wall of the tent and had crossed over and was in the act of coming down the other side when Bill burst in, his eyes blurred with sleep, a lighted lamp in one hand and a gun in the other.

I never was so disappointed in my life, because it wasn't a Hydrophobic Skunk at all. It was a pack rat, sometimes called a trade rat, paying us a visit. The pack or trade rat is also a denizen of the Grand Cañon. He is about four times as big as an ordinary rat and has an appetite to correspond. He sometimes invades your camp and makes free with your things, but he never steals anything outright—he merely trades with you; hence his name. He totes off a side of meat or a bushel of meal and brings a cactus stalk in; or he will confiscate your saddlebags and leave you in exchange a nice dry chip. He is honest, but from what I can gather he never gets badly stuck on a deal.

Next morning at breakfast Johnny and Bill were doing a lot of laughing between them over something or other. But we had our revenge! About noon, as we were emerging at the head of the trail, we met one of the guides starting down with a couple that, for the sake of convenience, we had christened Clarence and Clarice. Shorty hailed us.

"How's everything down at the camp?" he inquired.

"Oh, all right!" replied Bill—"only there's a good many of them Hydrophoby Skunks round. We saw four of 'em last night."

Clarence and Clarice crossed startled glances, and it seemed to me that Clarice's cheek paled a trifle; or it may have been Clarence's cheek that paled. He bent forward and asked Shorty something, and as we departed full of joy and content we observed that Shorty was composing himself to unload that stock tale for tenderfeet. It made us very happy.

By common consent we had named them Clarence and Clarice on their arrival the day before. At first glance we decided they must have come from Back Bay, Boston—probably by way of Lenox, Newport and Palm Beach; if Harvard had been a coeducational institution we should have figured them as products of Cambridge. It was a shock to us all when we learned they really hailed from Omaha. They were nearly of a height and a breadth, and similar in complexion and general expression; and immediately after arriving they had appeared for the ride down the Bright Angel in riding suits that were identical in color, cut and effect—long-tailed, tight-buttoned coats; derby hats; stock collars; shiny top boots; cute little crops, and form-fitting riding trousers with those Bartlett pear extensions 'midships and aft—and the prevalent color was a soft, melting, misty gray, like a cow's breath on a frosty morning. Evidently they had both patronized the same tailor.

## An Indian Uprising

He was a wonder, that tailor. Using practically the same stage effects, he had, nevertheless, succeeded in making Clarence look feminine and Clarice look masculine. We had gone down to the rim to see them off. And when they passed us in all the gorgeousness of their city bridle-path regalia, enthroned on shaggy mules, behind a flock of tourists in nondescript yet appropriate attire, and convoyed by a cowboy who had no reverence in his soul for the good, the sweet and the beautiful, we felt—all of us—that if we never saw another thing we were amply repaid for our journey to Arizona.

The exactly opposite angle of this phenomenon was presented by a certain Eastern writer, a member, as I recall, of the Jersey City school of Wild West story writers, who went to Arizona about two years ago to see if the facts corresponded with his fiction; if not he would take steps to have the facts altered—I believe that was the idea. He reached the hotel at Grand Cañon in the early morning, hurried at once to his room and presently appeared attired for breakfast. Competent eyewitnesses gave me the full details. He wore a flannel shirt that was unbuttoned at the throat to allow his Adam's apple full sweep, a hunting coat, buckskin pants and high boots, and about his waist was a broad belt supporting on one side a large revolver—one of the automatic kind, which you start in to shooting by pulling the trigger merely and then have to throw a bucket of water on it to make it stop—and on the other side, as a counterpoise, was a buck-handled bowie knife such as was so universally not used by the early pioneers of our country.

As he crossed the lobby, jangling like a milk wagon, he created a pronounced impression upon all beholders. The hotel is managed by an able veteran of the hotel business, assisted by a charming and accomplished wife; it is patronized by scientists, scholars and cosmopolitans, who come from all parts of the world to see the Grand Cañon; and it is as up-to-the-minute in its appointments and service as though it fronted on Broadway, or Chestnut Street, or Pennsylvania Avenue.

Our hero careened across the intervening space. On reaching the dining room he snatched off his coat and, with a gesture that would have turned Hackett or Faversham as green with envy as a processed stringbean, flung it aside and prepared to enter. It was plain that he proposed to put on no airs before the simple children of the desert wilds. When in Rome be a Roman candle—that was his motto evidently. He would eat his antelope steak and his grizzly b'ar chuck in his shirtsleeves, the way Kit Carson and Old Man Bridger always did.

## Hopi Hooligan's Sorrows

The young woman who presides over the dining room met him at the door. In the cool, clarified accents of a Wellesley graduate, which she is, she invited him to have on his things if he didn't mind. She also offered to take care of his hardware for him while he was eating. He consented to put his coat back on, but he clung to his weapons—there was no telling when the Indians might start an uprising. Probably at the moment it would have deeply pained him to learn that the only Indian uprising reported in these parts in the last forty years was a carbuncle on the back of the neck of Uncle Hopi Hooligan, the gentle copper-colored floorwalker of the white-goods counter in the Hopi House, adjacent to the hotel!

However, he stayed on long enough to discover that even this far west ordinary human garments make a most excellent protective covering for the stranger. Many of the tourists do not do this. They arrive in the morning, take a hurried look at the Cañon, mail a few postal cards, buy a Navajo blanket or two and are out

162

again that night. Yet they could stay on for a month and make every hour count! To begin with, there is the Cañon, worth a week of anybody's undivided attention. Within easy reach are the Painted Desert and the Petrified Forests—thousands of acres of trees turned to solid agate. If these things were in Europe they would be studded thick with hotels and Americans by the thousand would flock across the seas to look at them. There are cliff-dwellers' ruins older than ancient Babylon and much less expensive.

The reservations of the Hopis and the Navajos, most distinctive of all the Southern tribes, are handy, while all about stretches a big Government reserve full of natural wonders and unnatural ones too—everything on earth except a Lover's Leap. There are unexcelled facilities for Lover's Leaps too—thousands of appropriate places are within easy walking distance of the hotel; but no lover ever yet cared to leap where he would have to drop five or six thousand feet before he landed. He'd be such a mussy lover; no satisfaction to himself then—or to the undertaker either.

However, as I was saying, most of the tourists run in on the morning train and out again on the evening train. To this breed belonged a youth who dropped in during our stay; I think he must have followed the crowd in. As he came out from breakfast I chanced to be standing on the side veranda and I presume he mistook me for one of the hired help. This mistake has occurred before when I was stopping at hotels.

"My friend," he said to me in the patronizing voice of an experienced traveler, "is there anything interesting to see round here at this time of day?"

Either he had not heard there was a Grand Cañon going on regularly in that vicinity or he may have thought it was open only for matinées and evenings. So I took him by the hand and led him over to the curio store and let him look at the Mexican drawn-work. It seemed to satisfy him too—until by chance he glanced out of a window and discovered that the Cañon was in the nature of a continuous performance.

The same week there arrived a party of six or eight Easterners who yearned to see some of those real genuine Wild Western characters such as they had met so often in a film. They trotted out a troupe of trail guides for them—all ex-cowboys; but they, being merely half a dozen sunburned, quiet youths in overalls, did not fill the bill at all. The manager hated to have his guests depart disappointed. Privately he called his room clerk aside and told him the situation and the room clerk offered to oblige.

The room clerk had come from Ohio two years before and was a mighty accommodating young fellow. He slipped across to the curio store and put on a big hat and some large silver spurs and a pair of leather chaps made by one of the most reliable mail-order houses in this country. Thus caparisoned, he mounted a pony and came charging across the lawn, uttering wild ki-yis and quirting his mount at every jump. He steered right up the steps to the porch where the delighted Easterners were assembled, and then he yanked the pony back on his haunches and held him there with one hand while with the other he rolled a brown-paper cigarette — which was a trick he had learned in a high-school frat at Cincinnati — and altogether he was the picture of a regular moving-picture cowboy and gave general satisfaction.

If the cowboys are disappointing in their outward aspect, however, Captain Jim Hance is not. The captain is the official prevaricator of the Grand Cañon. He moons round from spot to spot, romancing as he goes.

## The Damaged Bride

Two of the captain's standbys have been advertised to the world. One of them deals with the sad fate of his bride, who on her honeymoon fell off into the Cañon and lodged on a rim three hundred feet below. "I was two days gettin' down to the poor little thing," he tells you, "and then I seen both her hind legs was broke." Here the captain invariably pauses and looks out musingly across the Cañon until the sucker tourist bites with an impatient "What happened then?" "Oh, I knew she wouldn't be no use to me any more as a bride — so I shot her!" The other tale he saves up until some tenderfoot notices the succession of blazes upon the treetrunks along one of the forest trails and wants to know what made those peculiar marks upon the bark all at the same height from the earth. Captain Hance explains that he himself did it — with his elbows and knees — while fleeing from a war party of Apaches.

His newest one, though—the one he is featuring this year—is, in the opinion of competent judges, the gem of the Hance collection. It concerns the fate of one Total Loss Watkins, an old and devoted friend of the captain. As a preliminary he leads a group of wide-eared, doe-eyed victims to the rim of the Cañon. "Right here," he says sorrowfully, "was where poor old Total slipped off one day. It's two thousand feet to the first ledge and we thought he was a gone fawnskin, sure! But he had on rubber boots, and he had the presence of mind to light standing up. He bounced up and down for two days and nights without stoppin', and then we had to get a wingshot to kill him in order to keep him from starvin' to death."

The next stop will be Southern California, the Land of Perpetual Sunshine—except when it rains!

CHAPTER NINE

# 1922
# Into the Grand Canyon,
# and Out Again, by Airplane

A. Gaylord

Wilderness enthusiasts are frequently offended by airplanes that fly over roadless areas and violate their sense of isolation. In spite of sophisticated legal definitions, wilderness is as much a state of mind as it is a place, and a state of mind is a very fragile thing. It is vulnerable to the *sounds* as well as to the sights of civilization. It can be an almost wrenching experience to labor for days to achieve geographical isolation and then have the sense of solitude shattered by the roar of an aircraft engine, whether one is disturbed simply by the noise or, less consciously, by the noise's message that isolation is practically impossible today anyway.

And so everyone has not viewed air travel in the Grand Canyon as an unmitigated blessing. This is no place for an examination of such an issue, for the questions and issues are many ("An injured hiker isn't nearly so bothered by the sound of a rescue helicopter," or "Flying over this place is like using a trampoline to get closer to the ceiling of the Sistine Chapel! If God had meant for us to see it from up there he'd have built a trail."). It is not inappropriate, however, to suggest that Mr.

167

Thomas's technological triumph offers fertile fields for thought.

Coincidentally, there appeared next to this article in *The Literary Digest* a testimonial ad for the "Remington Portable Typewriter," in which it was reported that members of an unsuccessful Mount Everest Expedition used the "writing machine" under "conditions where man could hardly live." Though adventurers have not lost their love of gadgetry since the 1920s, they would find some amusement, and maybe some similarities, in both the typewriter on Mount Everest and Mr. Thomas's airplane in the Grand Canyon.

A COMMANDER of the British Royal Flying Corps visited the Grand Canyon some twelve months ago and gave it as his opinion that landing in this great terrestrial crater would be extremely dangerous for an aviator because of the many treacherous air currents, and that the feat would probably not be attempted for some time to come.

Mountains, canyons, cliffs, rocks and trees, ravines and valleys disturb the air currents almost exactly the same as water is affected when flowing over and around huge boulders, cliffs and other obstacles, and the air has its eddies, whirlpools, up currents and down currents; its airfalls like the waterfalls.

Most aviators, I venture to say, would have been content to fly down into the canyon and make a safe landing. But not so Thomas. He was not satisfied with his performance until he had climbed back up again without landing and then dropped over the rim in a long tail-spin, which carried him nearly to the bottom, five thousand feet below, while throngs of tourists stood along the rim above and others astride donkeys paused on the steep trails below—gazing in open-mouthed astonishment. It was upon one of the plateaux that Thomas landed in his Thomas Special airplane.

Leaving his plane at Williams, Ariz., some days before his flight, he went by train to the canyon to inspect the valley for a possible landing-place. Arrived at the canyon he joined a party of tourists going down—on donkeys—and went with them down the Bright Angel trail. At Indian Gardens Thomas left the tourists and struck out across the valley and soon found a likely looking spot. It was covered with greasewood—a small western shrub, about eighteen inches high—and was fairly level and about 60x450 feet.

The following day he obtained permission from Colonel Crosby, park superintendent, to attempt the landing, and the latter sent a park ranger down with Thomas to mark the landing-place and to estimate the time necessary to clear and put it in shape for a landing. This was accomplished by five men in one day.

This done, Thomas returned to Williams, sixty-three miles distant from the canyon, and on the following morning, Tuesday, hopped off. Climbing to an altitude of one thousand feet, he pointed his plane toward the canyon, following the railroad. When within about eighteen or twenty miles of the big chasm he rose to an altitude of two thousand feet, for below him was thick shrubbery and trees, making it impossible to land without a crash, and altitude was necessary to permit him to make a long glide over this bad stretch back to safety in case anything went wrong with his motor.

Continuing at an altitude of two thousand feet until he reached the rim, Thomas circled out over the canyon to test the air, returning in a few minutes and dropping to within a hundred feet of the rim to take some pictures of El Tovar hotel and its cluster of small buildings, including Bright Angel cottages. Then, circling back over the railroad station, he dropped to about twenty feet above the buildings and the crowd of tourists, and to show the perfect balance of his plane, held up both hands, smiling as he glided by overhead.

Completing another circle and again flying low over the tourists standing along the rim, he headed straight for the edge of the big cut. To quote further from Mr. Gaylord's description:

The motor was ticking as steadily as a clock. Up to the rim, and then, with a throttled motor, he dropped slowly over and down—down into the very bowels of the earth!

The plane rocks a bit as it strikes an angry cross-current of air. Far, far below are rocks, rocks, rocks, and at the very bottom a silvery thread—the Colorado. Bright Angel trail creeps slowly up under the nose of the plane; then passes as slowly up and back behind, twisting and winding back and forth until lost from sight at the rim of this Devil's Bowl. Thomas looks over his shoulder and smiles—he is thinking of the many long hours he spent riding up and down, or rather down and up, that same awe-inspiring trail on the back of a donkey.

A group of pigmies on toy donkeys steals gradually into view under the nose of the plane. A wave of the hand and an instant later they are looking down upon the airplane—it is far below them.

Indian Gardens creeping up slowly under the plane; now the watering-place—and the plane passes out between Hopi point and Mojave point and into the great plateau section of the canyon. Below is the tiny landing-place—a small, flat, oblong, almost surrounded by rocks, pinnacles, towers and buttes. It is a simple matter now to make a safe landing. But does he do it? No. He is pointing the nose of the plane upward now and begins climbing in wide, graceful circles. He soon reaches an altitude of about four thousand feet from the bottom—still a thousand feet below the surface of the earth.

The motor slows down. Thomas waves his hand to the people gathered along the rim high above him. The nose of the plane shoots up. One wing drops. Then the nose topples over and the plane shoots down. The tail wiggles and twists. Down, down, down; five hundred feet, eight hundred feet, one thousand feet— the plane is plunging and whirling to the bottom at a terrifying speed.

Suddenly the motor begins to roar again. The plane has straightened out and now is flying on a level course. The most dangerous and yet the most useful stunt known to aviators has been executed for the first time in the very bowels of the earth!

The huge, graceful eagle turns slowly, circling gradually downward and, with diminishing speed, glides toward the small landing-spot among the boulders and gently, very gently, settles down and stops, rolling all the way across the small landing-spot and stopping about fifty feet from the edge of an 1,800-foot gorge.

Now, getting down into this rock-studded valley safely with an airplane was one thing; taking the air again from such a small clearing for the return trip back to the rim was another and entirely different matter. To take off, an airplane must get a fairly long run to pick up speed.

The return was not made until the following day, Wednesday. After landing and making the plane fast as best he could, Thomas returned to the rim, via the donkey route. He had no more than reached the top when word came to him that a high wind had turned his plane half-way around, breaking off the tail skid. This he repaired with a piece of broken automobile spring and wire.

At 10:12 o'clock Wednesday morning Thomas hopped off from the small plateau at the bottom of the canyon. The wind changed 60 degrees while the airplane circled this small plot once, and only a very short run could be made for the take-off, so that by far the most difficult phase of his undertaking was before Thomas. Indeed, it was this return flight to the rim that worried Colonel Crosby, the park superintendent, more than anything else.

Getting a short but fairly good run for it, Thomas banked the plane steeply against the wind and began to climb in very small circles, keeping always within easy gliding distance of the landing-spot. After circling the small field in this manner until the plane gathered speed he gradually widened the circles, reaching the rim at 10:17. Thus it required five minutes to climb the mile from bottom to top.

CHAPTER TEN

# 1941
# The Eighteenth Expedition

## Weldon Heald

Weldon Heald, a professional architect and avid mountaineer, made this trip down the Colorado River less than twenty years after it was said that boating the canyon "is not a method which has proved, or is likely to prove, popular."[1] In a way, we might think of his trip, "the eighteenth expedition," as the "last" expedition. Norman Nevills was the first commercial river runner, and this 1941 trip was his third through the canyon. After commercial river running was established, the word "expedition" became a little less accurate description of the trip. It did not suddenly become less dangerous, nor did discoveries cease, but later river runners have made it all seem almost commonplace by their very numbers. By 1967, 2,000 people were making the trip annually, and in 1972 a record 16,400 ran the river. After that the National Park Service found it necessary to place a ceiling on annual travel of about 14,000.[2]

But in 1941, only 17 parties had gone through, and several had met with tragedy. Heald describes the trip at a time when Grand Canyon river travel was still poised on the brink of modernity; when wild river rafting had neither the following nor the fashionable currency it would later earn.

Reprinted from *The Inverted Mountains* edited by Roderick Peattie by permission of the publisher, The Vanguard Press, Inc. Copyright 1948; Copyright © renewed 1975 by Mrs. Ruth Peattie.

NORMAN Nevills lives up on the San Juan River, a tributary of the Colorado, at Mexican Hat, Utah, where he has a private navy of fifteen boats. Boats are Norm's life, rivers his religion. They are in his blood. He has run six thousand rough-water miles on Western rivers with the record of never upsetting a boat.

But Norm's great love is the Colorado. No man knows it better. He has studied the river for twenty years, read, countless times, all that has been written about it, and shot every one of the thousand man-eating rapids of the Green and Colorado from Wyoming to Lake Mead in high water and low.

I first met Norm Nevills by chance in the spring of 1941 under the towering stone arch of Rainbow Bridge. Our party had packed in over the trail from Arizona while Norm had hiked up from the Colorado with some people he was piloting down to Lees Ferry. We talked for ten minutes, exchanged addresses, and went our ways. But I plodded up the trail nursing secret ambitions. Norm's river talk is infectious.

A month later I received a characteristic telegram from him. "Decided to shoot the Colorado a third time," it read. "Want another member to fill out crew and share expenses. This is definitely not a 'dude' trip, but to get pictures and work a kayak through for the first time. Will be a grand trip. How about it?"

I wired to Mexican Hat, "Can't row, but when do we start?"

So two months later we pushed off from Lees Ferry on our 350-mile journey to Hoover Dam. There were six in the party. Norm, with Agnes Albert of San Mateo, California, and William J. Schukraft from Chicago, led in the flagship, *Wen*. Dell Reed, Norm's neighbor from Bluff, Utah, and I followed in the *Mexican Hat II*. Our third boat, the kayak *Escalante*, was piloted by Alexander E. Grant from Dedham, Massachusetts. "W. E. N." were Norm's father's initials, but the "II" attached to our boat looked ominous. I never dared ask what happened to the *Mexican Hat I*.

According to the skipper's log, Nevills' 1941 expedition, the eighteenth successful party to make the passage of Marble and Grand canyons, departed from Lees Ferry at 12:15 P.M., July 15. He wrote:

These embarkations are awfully nervewracking, and it is with relief that I finally give the signal to shove off. The two fifteen-foot cataract boats are quickly readied for the start, but the kayak needs a lot of tinkering. As I look at the frail foldboat, I, like the others, have serious misgivings as to any possibility of its surviving the heavy water of Grand and Marble canyons. At any rate we will be able to pull Zee out even if we can't save the foldboat. So I'm hoping for the best.

We all were. Zee Grant's fragile entry in the Colorado River steeplechase was a sleek, cloth-covered craft, seventeen feet long, weighing but sixty pounds. The interior, with the exception of the

Norman Nevills, leaning on the gunnels of *WEN. Courtesy of the Grand Canyon Natural History Association*

cockpit, carried a varied assortment of inflated beach balls while sponsons made of huge inner tubes from Fifth Avenue bus tires were strung along each side to give steadiness. Zee carried a tire pump, for there are no service stations along the Colorado.

The rest of us were very kind to Zee before the start. We regarded him as a brave young man who had picked on the wrong river. However, Zee was no novice in rough water. He had won most of the foldboat championships in America. But the kayak's first test would come at Badger Creek, eight miles below Lees Ferry. Would Zee make it? We thought we knew the answer. Says Norm:

1:35 P.M.-3:50 P.M. BADGER CREEK RAPID. We pull in on the left and under the very slim shade of a great boulder have lunch. As soon as possible I go down to look over the rapid. My first hunch is confirmed; it can be run! This is a great satisfaction as in '38 I lined and last year there was a mighty tough and rocky channel. Now it is a straight shot through, guarded by two holes where the water plunges over submerged rocks. But there is a well-defined tongue. As it is hard to see from above I have Dell signal me for position so I won't miss the tongue. All is set. I slide through with the *Wen*, soon followed by the *Mexican Hat II*. I leave Dell below the rapid to bail and be ready to push out in case of an accident to the foldboat, and go up to signal Zee. He starts. Too far left. I signal him over. Now he's too far right! I signal him over. Still too far to the right. I signal him again. This time he whirls his boat about with a few heavy paddles—and he's 'way too far left. I yell and signal frantically. Zee raises up in the boat and sees his predicament, turns the foldboat around, paddles like a madman—but is swept on into the worst hole in Badger. As I start to run for the *Wen*, some two hundred yards downstream, I watch the river. The foldboat almost immediately sticks its nose through the big wave at the lower edge of the hole—upside down! No sign of Zee. The boat goes on through maybe a couple of hundred feet when I spot Zee bobbing up ahead of the *Escalante*! He holds himself back and grabs the boat. Dell has by now made it out into the river. We start to take off, but as I see that Dell has got Zee okay I pull back to shore. Later Zee explains that his air inflated life preserver went to pieces in the heavy water and all that saved him was an emergency gas bottle preserver! From the time I shoved off in the *Wen* until Zee got to shore downstream a ways, only twenty-five minutes elapsed.

But it seemed hours. I feel a great relief that the *Escalante* can get into such tough water and come through in such good shape. Zee never let loose his paddle during all the underwater maneuvers.

4:15-4:30 P.M. Here at Mile 8½ is where Zee landed on left side. We all assemble here to talk it over.

Zee described being sucked down, mauled, and rolled over and over, traveling twenty-five miles an hour under water. "I felt like a handkerchief in a washing machine," he said.

5:05-6:00 P.M. SOAP CREEK RAPIDS. We land on right, and a glance suffices to show that it is easily run. Although it is getting late we all want to take a crack at it. I think it's good psychology for Zee to take it on. I slip through, riding the big waves. They are *big* fellows. Good twenty feet high. I say to myself in the middle of them, "Why, they're just like mountains." Grand ride. Dell slips off to the right and misses the fun. Zee comes through riding the big ones high, wide, and handsome! The boy can really take it! My respect for the foldboat is rapidly mounting, though I do know that fast-hitting side waves or breaking waves will flip him. No good spot to camp so we go on.

6:15 P.M.-8:25 A.M. CAMP. Nice place. Left-hand side just below Canyon at Mile 12½. Eleven and a half mile run today.

This is the most we've made the first day out. We're all dog tired, but spirits are high.

JULY 16. EMBARK at 8:25 A.M. We are all eager to be on our way. I cook breakfast; the dishes are done by Dell.

8:50 A.M. SHEERWALL RAPID. Lots of fun. Agnes gets splashed.

Poor Brown. His dreams of railroad building ended here.

9:15-9:40 A.M. HOUSE ROCK RAPID. I have an idea this may be a tough baby, but upon drifting down upon it I see that one can easily slip off on the right side of the tongue to miss a few of the rather heavy waves in the main channel. I go ahead, signal, and Agnes and I run on through, followed by the *Escalante*. Zee does not back paddle enough and goes right on into the heavy water. About the third wave proves his nemesis, and over he goes! He climbs right back on the upside-down boat and paddles to shore without assistance. The *Mexican Hat II* comes through okay, and Zee's boat is soon ready to take off again. An upset a day! Oh me, what is this going to mean before we get to Lake Mead!

10:25-10:55 A.M. NORTH CANYON RAPIDS. I run through with the *Wen*, taking most of the big water. Zee takes the big water all the way—a very beautiful run. Dell slides off and misses the big fellows.

11:05-11:10 A.M. TWENTY-ONE MILE RAPIDS. We get a real ride, though it's not hard to run.

11:45 A.M.-12:30 P.M. MILE TWENTY-FOUR AND A HALF RAPID. *Wen*, *MH2*, *Kayak*, all through in fine shape. This is a bearcat in real low water.

12:40-2:15 P.M. LUNCH. Left. Good shade just above CAVE SPRING RAPIDS. In the cove here, stuck in a crevice, is an old handleless pick.

2:45 P.M. MILE THIRTY RAPID. This little rapid has a lot of whirlpools at the bottom, and we are much amused to see the *Escalante* spun around three times in short order.

4:00 P.M.-8:50 A.M. VASEY'S PARADISE. CAMP. Hard to land at this stage of water with very poor anchorage. We finally contrive to moor the boats, then all indulge in the luxury of the fresh, cold water from the waterfall and springs which are running more water than I've seen on the other two trips.

JULY 17. EMBARK, 8:50 A.M. This morning I am taking over the fold-boat for a try at it, while Bill and Zee are going to manage the *Wen*. At MILE 34 I run a little rapid in the *Escalante* and get through fine, though I find turning the foldboat much harder than the big boats. The trick seems to be to quarter the kayak away from what you want to miss, then paddle like the dickens. I take Agnes aboard with me! We arouse some consternation when I announce that she and I are going to run the next rapid together in the foldboat. We do so and come through with no more than a ducking. Then I turn the *Escalante* over to Agnes and get back aboard the *Wen*.

10:05 A.M. Arrive MILE 36½. I pull in left, followed by *MH2*. Agnes comes along drifting well out to midstream. I yell at her to make for shore as there is a heavy current into this rapid. Her efforts are not enough, and just by a miracle she gets the *Escalante* close enough to shore at the very brink of the rapid to get hung up on a rock. Dell takes me out past the rock in the *MH2*. I dive over to Agnes and get a firm grip on the foldboat while Dell goes on down over the rapid. A rope from shore is thrown us and we soon get landed.

10:25-10:55 A.M. I shove off in the kayak and run my first snorter in it. Make it fine, riding the big eddy right back to near the top, then run it again! All in all I run this one three times! Dell brings Zee through in the *Wen*, but she seems to feel a stranger at the oars as it nearly capsizes with Dell and he gets it full of water. Zee looks much perturbed.

12:02-2:30 P.M. We land on the right in the shelter of the upper of the ROYAL ARCHES. Good spot for lunch. It's windy on the river. Everybody sleeps as if we have nothing to do.

2:55 P.M. PRESIDENT HARDING RAPID. We arrive in a high, disagreeable wind. The rapid, as usual in high water, has quite a zip to it. We run on the left without stopping, though Zee lands for a second to get a better view.

3:40 P.M. Oh me! Wind and rain!

5:00 P.M. NANKOWEAP CREEK. CAMP. 20-mile run. After dinner tonight we light a great pile of driftwood that lights up the whole canyon. It sends a glow to the three-thousand-foot rim above and a great ways up and down the canyon. All in all it's a great camp. Wind has died down, and the weather looks pretty good.

JULY 18. We stay in camp today as we are ahead of schedule. This morning is devoted to a climb up to some Indian cliff ruins five hundred feet above the river and some others at the mouth of Nankoweap creek. Get back at 1:00 P.M., and the rest of the day is spent just lazying around talking about the big rapids that got away, etc. Yesterday Bill got quite a thrill out of a little rapid above MILE 36. He was much perturbed but came through grinning from ear to ear.

Just for fun I have put two X's on the foldboat to indicate the two upsets so far. Zee joins in with the idea gleefully.

Funny. After breakfast today we all took salt tablets, and Bill and I were positively sick. We almost lost our breakfast.

Tonight we have another big fire that is terribly spectacular and sit around talking until midnight. How time flies! We have lots of fun working ourselves into a lather over Hance, Sockdolager, and Grapevine—the three biggest rapids. I really believe that my psychology will work as usual in painting such an awful picture that the reality will seem much lesser.

JULY 19. We get up at 6:45 A.M. after a night of wind and sand blowing. Everyone seems happy and rarin' to go. Sky is overcast and maybe holds a storm for this afternoon.

SHOVE OFF 8:30 A.M. The NANKOWEAP RAPIDS are just fun.

9:25-9:35 A.M. Foot of KWAGUNT RAPID. Splashy and fun. No need to look over. *Wen* and *Escalante* come through without shipping water, but *MH2* gets a ducking. We stop to let Bill dry his movie camera.

10:15 A.M. Lots of fun and nice going. Here we enter Grand Canyon.

10:35 A.M. LITTLE COLORADO RIVER. Small flow in little Colorado Canyon, and I win a milkshake. We go right on as this is an uninviting spot.

11:25 A.M. LAVA CANYON RAPID. Nice and splashy. Have trouble through here in keeping bearings, as at different stages of water the rapids come and go, while the topography is confusing.

LUNCH. We go a few hundred feet over to Tanner's old camp at foot of disused trail for lunch. Left side. After lunch we elect to explore the old workings. At the mine entrance the exploring party dwindles down to Agnes and me. We go way back to every tunnel end and have much fun making spooky noises in the dark. There are several hundred feet of workings, in spots caving quite badly. In one place I go across an old plank that spans a water-filled shaft. It's quite a thrill. Upon returning to the mouth of the mine we all sit around awhile and talk. Return to Tanner's Camp, pack up the lunch equipment, and take off.

3:05 P.M.-7:50 A.M. MILE 68½ CAMP. Left. 16½ mile run. We have the same camp as in '38, even using the same fireplace. We go over to sand flat to build a big woodpile for a signal fire for watchers at Desert View on the South Rim. Get in for some heavy work! After dinner we go over to burn the pile. It makes a beautiful blaze, and we watch anxiously for answering lights, but we see none. Next year will arrange to have a rocket set off as a signal that we are observed. I later found that we were seen coming into camp this afternoon. We entertain ourselves with gruesome stories of the horrors awaiting us tomorrow at Hance, Sockdolager, and Grapevine.

There is a tingling feeling of suspense about running big rapids—half anticipation, half dread—which was intensified that first morning in the Grand Canyon by Norm's hair-raising stories of what was ahead. He keyed us up as a football coach does his team before the big game. Norm's method is pretty strong medicine. It even works on himself. But I suspect this watch-spring tension is the ideal attitude with which to run the Colorado. With it—danger can be fun.

JULY 20. Up at 6:00 A.M. Everyone in fine fettle and anxious for the thrill promised in today's run to Bright Angel Creek.

SHOVE OFF 7:50 A.M. TANNER RAPIDS 7:56 A.M. Lots of fun.

8:35 A.M. UNKAR CREEK RAPID. Land right. Mile 72½ We look it over and find it will be fun, but must use a bit of care. Away we go and have a marvelous ride.

Unkar Creek Rapid can be seen from Cape Royal on the North Rim, looking from five thousand feet above like a little riffle in a brook.

9:30-9:35 A.M. SEVENTY-FIVE MILE RAPID. A glance and away we go. Only trick to it is that you have to pull hard right at the bottom, and I don't get over far enough. We land in a hole, and what a ducking! Weldon out front swallows a ton of water. I have taken him over in my boat with Agnes, ready for the big fellows. This will leave Dell with only Bill.

To enliven things further I have a new song:
(Tune—A Man Comes To Our House)

Grapenuts in the morning,
Grapevine for lunch,
But I have a feeling, in fact it's quite a hunch:
When old man Hance sees us walking on the beach,
He is a mighty lesson, to us a'going to teach.

9:50-11:05 A.M. HANCE RAPID. Mile 76½ left. This is a whizzer. We look it over until 10:15, at which time I shove off in the *Wen*, running toward the left from the middle. I go through three holes in succession and fill up to the seat. What a ride! Pull in some ways below, bail, then signal Dell through. Comes through fine, then *Escalante* runs it, playing the left bank. This rapid is a number one toughy at any stage of water.

181

SOCKDOLAGER! 11:27-11:40 A.M. At top looking it over. 12:07 P.M. all safe at bottom. Well, this is a honey! After my big build-up I'm even surprised myself. In '38 this looked like a pretty formidable piece of water—but this year—oh me! The tongue drives hard to the right wall, bad holes on both sides, making a tricky problem to get through. We had a real thrill taking off. Made it fine. Foldboat and *MH2* come through beautifully.

GRAPEVINE RAPIDS. Mile 81½. 12:35-12:45 P.M. looks over. 1:00 P.M. all safe at bottom. After Sockdolager and the sight and ride it presented I didn't seem to feel quite so impressed with this fellow, although it has a mean channel. I took off with Agnes and Weldon, drifted down on the tongue, saw we were going too far right, so pulled hard left, and almost pulled us into some really tough holes. I pulled back hard to the right—and we had it made. For a minute, though, I thought church was really going to let out! We had promised ourselves not to eat lunch until Grapevine was under our belt, so we are looking for a lunch spot.

1:15-3:15 P.M. LUNCH. Left. Mouth of Boulder Canyon. Putting the toughies safely behind goes to our heads, and the first thing I know the whole caboodle of us are talking at the top of our voices at once!

3:35 P.M. EIGHTY-THREE MILE RAPID. It's nice running, though the waves are high and sharp. This is grand going from here on in, and we're riding them high, wide, and handsome. We go out of our way in the *Wen* to get a ducking—and get it! I get off my bearings a bit and keep thinking each rapid we come to is MILE EIGHTY-THREE. There is some comment.

Just above Bright Angel we are horseplaying. Agnes falls off the stern deck on top of me. I lose my balance and collapse on Weldon in the bow! What a day, and what a grand bunch. This is truly the merriest and happiest outfit that ever tackled the Colorado. This crowd bears out my theory that one or two gripers or whisperers can spoil a whole party.

We have fortified ourselves for a gala reception at Bright Angel Trail. We feel sure there will be a clamoring multitude to welcome us.

4:15 P.M. BRIGHT ANGEL CREEK. Mile 87½. Woe is me! I really thought we would have at least some kind of a reception committee. But not a soul! Not a solitary person is in sight as we come majestically in, holding perfect formation for the photographers.

## "High, Wide, and Handsome"

Phantom Ranch lies at the bottom of the Grand Canyon close by clear, sparkling Bright Angel Creek. We spent two days there making the most of our one contact with civilization. The six of us reveled in the sybaritic pleasures of soft beds, cold drinks, showers, and a swimming pool.

At Phantom, Norm's wife, Doris, met us with mountainous stacks of provisions packed down the trail on mules from the South Rim.

"You can't run the Colorado on hardtack and beans," says Norm. And so we enjoyed such rare delicacies as fresh bread and eggs throughout the trip. Several hours were required to pack all the food and stow it away in the watertight hatches of the cataract boats. They were groggy from weight when we pushed off from the beach at Phantom Ranch on July 23.

Somehow it seemed as if the trip really began after leaving Bright Angel Creek. For ten days we would be isolated at the bottom of one of the world's great gorges with no way out except by shooting a hundred roaring rapids. But the six of us were finally a working unit ready for anything the Canyon might offer.

And we found, soon enough, that the Canyon meant business. At Hermit Falls, six miles below Phantom Ranch, we had our first lining job. There the river drops too fast even for spunky cataract boats to dare the twenty-five-foot waves. Lining is an arduous task, sometimes taking three hours before the last boat is through. But to us it was a new experience spiced with a different kind of danger.

In lining, everything movable in the boats must be unloaded and portaged to the foot of the rapid. Then, with every man pushing and pulling with all his strength, the six-hundred-pound boats are eased down over the boulders and cascades through the narrow channels close to the shore. As the boats haltingly progress they are snubbed, first with the bow line, then the stern line, to prevent them from being whisked away by the current or smashed against a rock.

Lining is slow work, but exciting. The channels are deep and swift, while the footing is slippery, and glasslike rock protrudes from the water at all angles. Fortunately for our aching muscles Lava Falls, a hundred miles below, was the only other rapid we were forced to line.

We were dead tired the first night out from Phantom Ranch. We had run Granite Falls and Horn Creek Rapid as well as lining Hermit. But a sandstorm came up, making our dreams restless affairs in which we continued to fight the river all night.

JULY 24. 8:10 A.M. SHOVE OFF. Run BOUCHER RAPID right off the reel. Wonderful fast going. Surprisingly enough it's about the fastest water we've had so far on the trip. We made a good twenty miles an hour.

8:50 A.M. MILE 99½. This stretch through here is real sport and goes like the dickens over these small rapids. Hardly a place in the canyon more fun to run than these.

9:30 A.M. MILE 104½. RUBY CANYON RAPIDS. Run all these just looking over from the boat. River a bit red this morning but doesn't seem to be any higher. In '40 we had quite a bit of grief through here with rocks, but this high water is a cinch and is perfect going. We get a few drops of rain.

10:05-10:15 A.M. Well! This rapid has definitely changed! In '38 the channel *was* a sort of serpentine affair, staying mostly to the right. But this time the takeoff is near the left and switches over to the right a bit, then drops on through. It's beginning to rain.

11:00 A.M.-1:14 P.M. SHINUMO CREEK. LUNCH. We are glad to pull in here, though it's just stopped raining. We are cold so the boys build a fire. Funny weather. We stop here quite a while and take naps. We are finally disturbed by the hot sun driving down on us!

Running rapids, loading, bailing, and landing boats are damp occupations. But being constantly wet had its practical side. A merciless summer sun beats down into the barren canyons, sending the midday temperatures skyrocketing to 110°-115°F. But as we were air-cooled by evaporation all day long none of us suffered with the heat. In fact, when the sun disappeared behind high-piled thunderclouds, which gathered on the Canyon rims each afternoon, we actually became cold.

1:35-2:00 P.M. Left under overhanging ledge. We pull in here, driven to shelter by a terrific upstream driving rain! We are thrilled by the sight of many waterfalls forming from the literal cloudburst. Cross the canyon rocks are falling from the cliffs. One on our side, as big as an army tank, plunges into the river close to us.

WALTHENBURG RAPIDS. Arrive 2:25 P.M. *Wen* off 3:00 P.M. All at bottom 3:25 P.M. Leave 3:45 P.M. Another surprise! A flood has overtaken us, and this rapid has strong lashing waves in the center of the main drag. I decide to run the main channel. This is tough water. The waves have an awfully hard lash to them. The *Escalante* sneaks on the right and plays the filth, hits two rocks, and in crossing gets some hard going over by heavy water.

4:35 P.M.-8:50 A.M. ELVES CHASM. CAMP. Mile 116½. 20-mile run today. Difficult landing but we anchor on left. The creek is running red-colored water. Creek is on left, then a hundred and fifty feet of sand offers a good camping place. The boats are tied up and I have just put out the camping equipment when I see figures madly dashing around. I join the rush and find that another fork of the creek has discharged a flood that is coming clear across the sand bank. We gather up all equipment in the nick of time and lose nothing. But most of the sand is washed away, leaving ledges. It's quite a thrill and novel experience. This second water is a dirty brown. Although the flood passes over the boat anchorage it holds fast. The flood hit at 6:15 P.M. Little sand left, but a grand night's sleep nevertheless.

JULY 25. SHOVE OFF 8:50 A.M. Well, well, well! In taking off I bang into a rock, bounce off, and hit another!

The river rose three feet. I almost forgot: yesterday, in climbing from his foldboat at Walthenburg Rapids, Zee went up the cliff some fifteen feet, lost his grip, and came tumbling down into the river. What luck that no rocks were lurking under the surface! This stream at Elves Chasm is known as Arch Creek. We leave a register up in the cave for future parties to sign. I am going to get the names of past parties gradually filled into it.

It's clouded up again this morning. Looks as if rain would overtake us once more.

10:00 A.M. MILE 121½ RAPID. The bow-heavy *Wen* really got us slugged in this one! Going to have to shift my load. The foul red water is unpleasant stuff to get socked with.

10:40 A.M. FOSSIL RAPIDS. Fun to run. Starting to rain and is getting cold.

11:15 A.M.-1:35 P.M. LUNCH. Build a fire here where we've pulled in on the right. See a king snake. It's banded black and white, 18 inches long. Finally stops raining.

BEDROCK RAPID. Arrive 2:35 P.M. At bottom 3:10 P.M. On again 3:25 P.M. A real kick and a dangerous piece of water. We have a bit of time getting Zee to a vantage point to look it over. Weldon goes overland to lighten Dell's boat and meets us at bottom. It's a hard one to climb around, he finds. Agnes and I shove off in *Wen*. Near bottom get a good ducking. Dell and Zee come through fine. Bill rides with Dell.

DEUBENDORFF RAPID. Arrive 3:45 P.M. *Wen* off 4:00 P.M. All at bottom 4:25 P.M. On again 4:35 P.M. This is a honey! I go through, get socked in a hole, blinded by muddy water. Dell right in my tracks. Zee plays the right bank and keeps out of the big water. Good thing, as the power in these waves would have thrown him over like lightning.

5:05-5:15 P.M. TAPEATS CREEK. Land on right looking for a camp spot, but this is definitely no good. No anchorage, a flood down Tapeats Creek would catch us, and there's quicksand here! Therefore we decide to drop down a ways.

5:20 P.M.-8:55 A.M. CAMP. MILE 133½. 17½ miles today. We are now camped on the big bar just below Tapeats Creek. It's like ice water, but most of the gang take baths. Looks stormy a bit, but don't think it will rain. Nice camp. The sunset lights back on the Powell Plateau are gorgeous.

JULY 26. Up this morning with just a few clouds. River up six inches to a foot.

8:55 A.M. SHOVE OFF. Fine going this morning.

9:30-9:50 A.M. DEER CREEK FALLS. Water over falls half again as big as in '40. Brrr! The air and water are cold!

10:05 A.M. Indian cliff ruins. See some new ones this trip just upstream. Don't stop though.

10:12 A.M. MILE 138. This is the famous rapid where I dumped Doris and John in '40.

10:29 A.M. FISHTAIL CANYON. Small rapid here.

11:15-11:20 A.M. KANAB RAPIDS. Glance over from right. Marvelous riding, and we hate to see them end.

12:05-2:20 P.M. Left under ledge. LUNCH. MILE 148½. We are just eating when a heavy rain comes up. We are really pleased as it has caught us in a grand sheltered spot. After a bit of rain some beautiful waterfalls begin to come over the cliff walls, and we are treated to a rare sight. The high, ribbon-like falls come down from around 1,600 feet. Sky clears and we have blue sky for take-off.

Just before we get to UPSET RAPIDS I shove the *Wen* under a big waterfall and try to get Agnes and Bill wet. Waterfalls everywhere pouring over the cliffs.

UPSET RAPIDS. 2:40-2:50 P.M. All at bottom 2:55 P.M. This would be a tough nut to walk around in this high water, and anyway we're all spoiling for a good ride. Off we go! Boy, oh boy! One big wave passed us up going like a freight engine. I swear if it had hit us we'd have stayed hit!

4:25-5:00 P.M. HAVASU CANYON. MILE 156⅔. This is a disappointment. Here I've extolled the beauties of the sky-blue water, and we find the rains have made a red stream instead. We row right up through the narrows to the first waterfall. We try for a camping spot, but although it's late we decide to look for a more favorable place. Remembering the high walls through this section I have qualms about picking a camp site this side of MILE 164 CANYON.

6:30 P.M.-10:00 A.M. Right MILE 164 CANYON. 30½ miles today. We came rolling in anxious to make camp. Into MILE 164 RAPID we went. It was rough. Current drives hard to the left wall, and there are some ugly waves. We were barely through when Dell and Weldon ploughed in. Dell later reported that they came very near to upsetting when a hard lashing wave struck them. But Zee—over he went! He climbed back in, found he was facing wrong, and turned around! We all landed laughing, completely exhilarated by our experience. Zee made it right on in under his own power. Nice camp here, and we like it. As we are ahead of schedule we plan to fiddle around in the morning, taking our time.

## The Last Hundred Miles

Easy days are few on the Colorado, but the twenty-seventh was one which gave us a rare opportunity to relax and enjoy the river. The rapids were just big enough to be interesting without the sense of strain which we had been under for the past five days. So we shouted companionably from boat to boat or took our ease on the decks, marveling at the ever changing procession of rock towers and pinnacles outlined against the blue Arizona sky.

A small excitement was logged at lunch:

> We go into the mouth of a canyon a ways and find shade under a tree. Partly through lunch I glance up and see a bull snake resting in a branch above Dell's head. This evokes a mad scramble. We shortly move on and take off again.

Animal life was surprisingly scarce. Each day we scanned the cliffs for mountain sheep. Signs were plentiful, and at one place we came across a fine skull with huge curving horns, but we never caught sight of those famous hoofed rock climbers. Of rattlesnakes, centipedes, and scorpions we saw no trace, although we kept our eyes open in stony and brushy places.

On the afternoon of July 27 we had our second lining job:

> LAVA FALLS. Arrive 2:40 P.M. Same tough setup as in '38. We obviously have to line. Start lining: *Wen* 3:05, finish 4:10 P.M.; *MH2* 4:30, finish 5:00 P.M. Zee lines *Escalante* alone.

July 28 was another leisurely day, as we were well ahead of schedule. The skipper recorded, "It's fun to have no place to go and lots of time to get there." So we camped in the middle of the morning.

> 10:50 A.M.-8:45 A.M. CAMP. Right. Nice willow tree and a dandy place to lay over. Mile 192½. After lunch Agnes and I take on a sixty-foot basalt cliff back of camp. I take one ten-foot spill and never do get anywhere. We all join in a good laugh as I'm supposed to be giving climbing lessons! Nothing daunted, Agnes and I set out to climb a peak that rises about a thousand feet. We take no water and reach the summit dry. Grand view. See signs of mountain sheep. Going down Agnes gets some cactus thorns! We land in camp *dry*! Some grapefruit juice really hits the spot. After dinner we keep awake for some time swapping yarns, singing, etc. We are anxious to get some fresh water as this river water is a bit flat. First quarter moon is overhead.

The next morning we saw the first signs of human life since leaving Phantom Ranch six days before.

10:05-10:35 A.M. PARASHONT WASH. MILE 198½. As we are going by I look toward a ledge at right of canyon. See something suspended by a wire. Possibly a camp. We land to investigate. It proves to be a trapper's camp. A coyote trap is sprung close by. A stack of dishes under an oil can. Around the corner more evidences of recent camping. Nobody around though.

Sixteen days from our start at Lees Ferry we came to our last camp at Diamond Creek. Hoover Dam was still over a hundred miles away, but for most of the distance we would be towed down the blue waters of Lake Mead.

We lay over another day at our final river camp, spending most of the time enjoying a busman's holiday immersed in a Diamond Creek pool christened "Lake Zee" in honor of its chief engineer.

At Diamond Creek civilization was just around the corner. Our river days of unbroken companionship, unfailing good humor, and perfect co-operation were soon to end. We all regretted it. We had grown to believe it was the natural order of things for the six of us to be shooting the Colorado together; what was abnormal was the world to which we were returning. Agnes best expressed the feelings of the entire party when, after saying her last good night, she called back to us above the roar of the river, "I'd turn around tomorrow and do it all over again."

At noon, August 1, the three boats slid down Bridge Canyon Rapid into the quiet water of Lake Mead.

Three days later Norm wrote his last entry in the log for the Eighteenth Expedition:

Around 9:00 A.M. we approach the BOULDER CITY LANDING. Met by big boats. Pictures taken, and Zee is sensation of the hour—and justifiably so as he turned in a swell job of bringing his foldboat through. Well, anyway we made it!

As we tied up to the landing a lady ran up to Norm. "Oh, Mr. Nevills," she said, "was there any time when you were really frightened?"

"Only once," Norm assured her.

"When was that?" she asked.

"From the time we shoved off until we reached Lake Mead."

189

# SOURCES OF THE READINGS

Burroughs, John: "Temples of the Grand Canyon" originally appeared as "The Grand Canyon of the Colorado" in *Century Illustrated Monthly Magazine,* January, 1911, pp. 425-38.

Cobb, Irvin S.: "Roughing It Deluxe" from *The Saturday Evening Post,* June 7 and 28, 1913.

Garland, Hamlin: "The Grand Canyon at Night" from C. A. Higgins et al., *The Grand Canyon of Arizona* (Santa Fe: The Passenger Department of the Santa Fe Railroad, 1906), pp. 61-62.

Gaylord, A.: "Into the Grand Canyon and Out Again, by Airplane" from *The Literary Digest,* October 7, 1922, pp. 63-64.

Grey, Zane: "An Appreciation of Grand Canyon" from John Kane, ed., *Picturesque America, Its Parks and Playgrounds* (New York: Resorts and Playgrounds of America, 1925), p. 125.

Heald, Weldon F.: "The Eighteenth Expedition" from Roderick Peattie, ed., *The Inverted Mountains* (New York: Vanguard Press, 1948), pp. 187-95.

Hogaboom, Winfield: "To the Grand Canyon on an Automobile" from *Los Angeles Herald Illustrated Magazine,* February 2, 1902, pp. 18-20.

Jordan, David Starr: "The Land of Patience" from C. A. Higgins et al., *The Grand Canyon of Arizona* (Santa Fe: The Passenger Department of the Santa Fe Railroad, 1906), p. 88.

McCutcheon, John T.: *Doing the Grand Canyon* (Kansas City, Mo.: Fred Harvey, 1909), 19 pp. This short book is reproduced here in its entirety.

Monroe, Harriet: "The Grand Canyon of the Colorado" from *Atlantic Monthly,* December, 1899, pp. 816-21.

Muir, John: "Our Grand Canyon" originally appeared as "The Grand Canyon of the Colorado" in *Century Illustrated Monthly Magazine,* November, 1902, pp. 107-16.

Powell, John Wesley: "First Through the Grand Canyon" from the book of the same name (New York: Outing Publishing Co., 1915), pp. 198-257.

Priestley, John Boynton: "Grand Canyon, Notes on an American Journey" from *Harper's Magazine,* February-March, 1935, excerpts from pp. 276, 399-400.

Roosevelt, Theodore: Speech given at the Grand Canyon, May 6, 1903, from the *New York Sun,* May 7, 1903.

Warner, Charles Dudley: "The Heart of the Desert" from *Our Italy* (New York: Harper & Brothers, 1891), pp. 177-200.

Wister, Owen: Foreword to Ellsworth Cobb, *Through the Grand Canyon from Wyoming to Mexico* (New York: MacMillan, 1914), excerpts from pp. viii-ix.

# NOTES

*Introduction*

1. John Muir, "The Wild Parks and Forest Reservations of the West," *Atlantic Monthly,* January 1898, p. 28.
2. Thomas Lounsbury, "Biographical Sketch," in *The Complete Writings of Charles Dudley Warner* (Hartford: The American Publishing Company, 1904), p. xx.

*Chapter 2 (1891: "The Heart of the Desert" by Charles Dudley Warner)*

1. Thomas Lounsbury, "Biographical Sketch," in *The Complete Writings of Charles Dudley Warner* (Hartford: The American Publishing Company, 1904) p. xx. The only one of Warner's books that still has wide appeal is *The Gilded Age,* written with his friend and neighbor, Mark Twain.
2. C. A. Higgins et al., *The Grand Canyon of Arizona* (Santa Fe: Passenger Department of the Santa Fe, 1906), p. 89.
3. J. Donald Hughes, *In the House of Stone and Light* (Grand Canyon: Grand Canyon Natural History Association, 1978), p. 49.

*Chapter 4 (1902: "To the Grand Canyon on an Automobile" by Winfield Hogaboom)*

1. Earl Pomeroy, *In Search of the Golden West* (New York: Alfred A. Knopf, 1957), p. 125.
2. *Coconino Sun* (Flagstaff, Arizona), January 25, 1902.

*Interlude: Grand Canyon Portfolio*

1. William Allen White, "On Bright Angel Trail," *McClure's Magazine,* September 1905, pp. 503, 515.
2. Hamlin Garland (1860-1940) was the author of *Main Traveled Roads, A Son of the Middle Border,* and *A Daughter of the Middle Border.* He won the Pulitzer Prize for his auto-biographical account of farm life in the middle west.
3. The Grand Canyon Forest Reserve had been created by President Harrison in 1893, but this did not adequately protect the area from development by mining and grazing in-terests: at least it did not do so well enough to satisfy the people who wanted the canyon to be preserved in its primitive state. By the time Roosevelt became president in 1901, there was growing interest in having the Grand Canyon declared a national park.
In 1906 Roosevelt gave it additional protection by proclaiming it a game reserve, but this was still not enough. Some active mining interests threatened to mar the landscape and monopolize the best overlooks. Local attitudes varied, but nearby communities objected to what they feared would be a land lockup if the national park was proclaimed. They thought their own best interests would be served by mineral development (at several parks, local interests have been slow to realize there is more gold and silver to be mined from tourists than from the land).
Also in 1906, in legislation which seemed far remote from the issue of the Grand Canyon, Congress passed the Act for the Preservation of American Antiquities. This act allowed the president to create "national monuments," by which writers meant "objects of historic or scientific interests." The intent was to allow the president to protect small his-toric sites and curiosities, but the act gave Roosevelt the opportunity he needed. Since he could not create a national park (only Congress can), he chose to interpret the Antiquities Act very broadly, and use it to accomplish the same thing. Of course, such a liberal use of what Congress had imagined to be a very restricted power caused great uproar. Roosevelt was challenged, and his action was eventually upheld by the Supreme Court. Problems with mining claims would continue for some years, but national monument status gave the

canyon the protection it needed until support could be marshalled for the final drive to make it a national park.

Ten years after he gave the speech included here, Roosevelt was in the canyon hunting cougars, but even in the heat and excitement of the chase he was compelled to pause in wonder:

> From the southernmost point of this table-land the view of the canyon left the beholder solemn with the sense of awe. At high noon, under the unveiled sun, every tremendous detail leaped in glory to the sight; yet in hue and shape the change was unceasing from moment to moment. When clouds swept the heavens, vast shadows were cast; but so vast was the canyon that these shadows seemed patches of gray and purple and umber. The dawn and the evening twilight were brooding mysteries over the dusk of the abyss; night shrouded the immensity but did not hide it; and to none of the sons of men is it given to tell of the wonder and splendor of sunrise and sunset in the Grand Canyon of the Colorado. (From "A Cougar Hunt on the Rim of the Grand Canyon," *The Outlook,* October 4, 1913, p. 266.)

4. Zane Grey, *Tales of Lonely Trails* (New York: Harper & Brothers, 1922), pp. 167-68.

*Chapter 7 (1909: "Temples of the Grand Canyon" by John Burroughs)*

1. Clara Barrus, "With John O'Birds and John O'Mountains in the Southwest," *Century Illustrated Monthly Magazine,* August 1910, p. 521. See also John Burroughs, *Camping and Tramping with Roosevelt* (Boston: Houghton, Mifflin and Co., 1907), and Clara Barrus, *The Life and Letters of John Burroughs* (New York: Houghton, Mifflin and Co., 1925), p. 48.

2. Clara Barrus, "With John O'Birds and John O'Mountains in the Southwest," p. 521.

3. Ibid.

*Chapter 10 (1941: "The Eighteenth Expedition" by Weldon Heald)*

1. Roderick Peattie, introduction to *The Inverted Mountains* (New York: Vanguard Press, 1948), gives biographical background on Heald. The quotation is from the editorial introduction to A. Gaylord's "Into the Grand Canyon, and Out Again, by Airplane," *The Literary Digest,* October 7, 1922, p. 63.

2. J. Donald Hughes, *In the House of Stone and Light* (Grand Canyon: Grand Canyon Natural History Association, 1978), p. 114.

# SUGGESTIONS FOR FURTHER READING

THE accounts included in this book contain many references to people, places, and events that do not get fully explained. Because the book's primary purpose is to revive the accounts themselves, it did not seem appropriate to also provide extensive annotation. If these occasional references to unidentified things have increased your interest in the Grand Canyon, all the better. You will find the answers to any questions raised by this book in the books listed below.

One other point should be mentioned. This is not a reference book, and so the information presented by the various authors should not be regarded as uniformly accurate. Remember, first, that many of the authors were popular journalists, and, second, that knowledge of the canyon has increased through the years, so that even if their "facts" were considered correct at the time they may have since been disproven. Again, the references that follow will provide up-to-date information of many kinds about the canyon.

The literature of the Grand Canyon is immense, and I am tempted to begin listing other favorites of mine that, for one reason or another, were not included here. But the list would quickly grow too long, so I offer instead some general guides.

There are several reasonably recent histories of the Grand Canyon, including the following: C. Gregory Crampton, *Land of Living Rock* (New York: Alfred Knopf, 1972), J. Donald Hughes, *In the House of Stone and Light* (Grand Canyon: Grand Canyon Natural History Association, 1978), Joseph Wood Krutch, *Grand Canyon, Today and All Its Yesterdays* (New York: William Sloane Associates, 1958), T. H. Watkins, general editor, *The Grand Colorado, The Story of a River And Its Canyons* (Palo Alto: American West Publishing Company, 1969), and Roderick Peattie, editor, *The Inverted Mountains: Canyons of the West* (New York: The Vanguard Press, Inc., 1948). I am sure there are others equally deserving of mention, but this list is not intended to be comprehensive. The ones mentioned cover history, natural history, and lore, and do so in a variety of ways.

194

Another superb collection of writings about the Grand Canyon, edited by Governor Bruce Babbitt of Arizona, is *Grand Canyon: An Anthology* (Flagstaff: Northland Press, 1979).

Though some of the above books have excellent bibliographies, I also found the following useful in my search for early accounts of the Grand Canyon area: Anon., "Bibliography of Books, Government Reports, and Magazine Articles on Grand Canyon National Park," in *Report of the Director of the National Park Service to the Secretary of the Interior for the Fiscal Year Ending June 30, 1919* (Washington, D.C.: U.S. Government Printing Office, 1919), Anon., *A Bibliography of National Parks and Monuments West of the Mississippi River,* Volume II (Washington, D.C.: National Park Service, 1941), and Francis Farguhar, *The Books of the Colorado River and the Grand Canyon* (Los Angeles: Glen Dawson, 1953).

I hope the present book starts you on further explorations of the literature of the Grand Canyon. The possibilities for adventure and discovery are surpassed only by those of the Canyon itself.

7388